AF552825

ENCYCLOPAEDIA OF DEVELOPMENTAL BIOLOGY-III

MAMMALIAN DEVELOPMENT

By

Manju Yadav

Lecturer
Department of Zoology
M.M.H. College
Ghaziabad (U.P.)
(India)

DISCOVERY PUBLISHING HOUSE PVT. LTD.
NEW DELHI-110 002

First Published – 2008

Reprinted – 2025

ISBN: 978-81-8356-299-7

Mammalian Development

Published by:

DISCOVERY PUBLISHING HOUSE
4383/4B, Ansari Road, Darya Ganj
New Delhi-110 002 (India)
Phone: +91-11-23279245; 23253475; 43596065
Mobile: +91 9811179893 / +91 9871656464
E-mail: discoverybooksindia@gmail.com
orderdphbooks@gmail.com
namitwasan9@gmail.com
web: www.discoverypublishinggroup.com

Printed at:
Infinity Imaging Systems
Delhi

simple commodity production: the unity of the labour process and the process of creating value; differentiated from capitalist production of commodities in that it is undertaken by independent commodity producers who exchange the products of their labour rather than their labour-power: the basis for the development of surplus-value, wage labour, is absent.

simple reproduction: the expenditure of all of surplus-value as revenue so that there is no capital accumulation.

socially necessary labour time: the average amount of labour expended in the production of a commodity under current conditions (including, for example, the average productivity and intensity of labour).

surplus labour: that part of the working day extending beyond necessary labour; the time during which the worker produces surplus-value for the capitalist.

surplus-value: value accruing to the capitalist as a consequence of the working day extending beyond necessary labour time; the rate of surplus-value, the ratio of surplus-value to variable capital (wages) or of surplus to necessary labour time, expresses the degree of exploitation of labour by capital.

use-value: the use(s) to which a commodity may be put; its utility based upon its concrete properties as an object.

value: the amount of socially necessary labour time for the production of a commodity incorporated in that commodity.

variable capitol: that portion of capital expended (in wages) for the purchase of labour-power; that part of advanced capi tal which expands its value in production.

wage: the value of labour-power in money expressed on the surface of society as the price of labour; in real terms, the sum of commodities comprising the worker's subsistence.

Preface

The present title **Mammalian Development** is an important link between almost all biological sciences, and is assuming more and more importance in fields such as medicine and agriculture. Molecular biology has began fulfilling its long anticipated role of linking genetics and embryology and many questions of embryology that had lain dormant for decades have been taken up with new molecular tools and strategies. While our knowledge of the molecular aspects of development has so drastically increased during the time, our applications of developmental phenomenon in evolution and ecology has also expanded. It is becoming impossible to study any area of biology without some background in developmental biology.

This title integrates the descriptive, experimental and biochemical approaches into a conceptual framework for the analysis of development. All important points are illustrated diagrammatically. It is well balanced, completely accessible presentation of principles and application of embryology. In order to create a book which is written within the economy as well as the physical group of the student, only the main essentials of each topic have been presented. It does this with the aid of careful selected examples—some recent and other classic of the field and with numerous illustrations. The aim is to enthuse the reader with this active and exciting area of research and to lay a solid foundation on which further study of its various facts may be based.

To make the work more comprehensive and informative, the author has consulted many authoritative books, research journals, abstracts, monographs etc. He is grateful to all those great scholars whose work are cited or substantially reproduced.

There can be no claim to originality except in the manner of treatment and much of the information has been obtained from the books and scientific journals available in the different libraries.

The author expresses his thanks to his friends and colleagues whose continue inspirations have initiated him to bring out this book.

The author expresses his gratitude to Mr. Wasan and staff of M/s Discovery Publishing House for their whole hearted co-operation in the publication of this book.

Author

CONTENTS

1

Introduction

One characteristic of living things is that they reproduce. This fact is so familiar that we are apt to lose sight of the wonder of it. But it is one of the astounding realities of nature; and it is the responsibility of embryologists to describe it and to try to explain how it is accomplished.

In a broad sense, embryology includes more than the study of the development of eggs. It belongs to the larger field of *developmental biology.* Some organisms grow from buds, or from fragments of parent organisms. Moreover, the regeneration of lost parts and the healing of wounds reawaken processes related to embryonic development and are therefore of concern to embryologists. Aging and rejuvenescence (if such there be) and cancer are aspects of development. In short, *embryology* is concerned with the origin of structure, in contrast to *physiology,* which studies the processes by which structure is maintained. The structure of living things, however, is not to be thought of as something static like that of a stick or stone. On the contrary, the structure of cells and tissues is sensitive and changing. It cannot be separated from function any more than function can exist apart from structure. It is maintained by a constant expenditure of energy. Hence the study of the origin of structure and function (morphogenesis) and the study of the maintenance of structure and function (homeostasis-an important aspect of physiology) cannot be wholly separated. When we study embryology, we see developmental processes in action which are never absent wherever and whenever life exists.

For many years embryologists were concerned with describing the stages of development: first as they saw it with their naked eyes; then with the help of magnifying lenses; and then with the aid of compound microscopes which magnify up to 1,000 times. Techniques were devised by which tissues could be hardened, sectioned, and stained for detailed study. As a result, today we have for some animals a rather full descriptive account of what happens at the tissue level and even at the cellular level when an organism develops. Other animals remain to be studied and many details need to be filled in, but the outlines are clear.

Mere description, however, does not explain. It does not tell us of the processes which are at work in time and space by which a new individual comes into being. To answer such questions, experiments have been undertaken on eggs and embryos. The eggs of certain marine invertebrates, as well as those of frogs, salamanders, and chicks, have proven especially useful. This work is progressing in many laboratories throughout the world at the present time. Early leads have sometimes proved false, but gradually the principles and processes which underlie development are coming to light.

In recent years, especially since the Second World War, electron microscopes have been devised which magnify as many as 200,000 times. These have provided new and exciting insights into cellular structure and activities. The observations have confirmed, for the most part, what the light microscope had already revealed, while at the same time extending our knowledge. Unfortunately, living material cannot be studied with the electron microscope. It must first be killed and usually sectioned before being photographed under the electron beam. By piecing together the information obtained with the light microscope about materials living and dead with what the electron microscope reveals, and by using all sorts of materials, a picture is being gained of what goes on within cells when development takes place.

Methods have been devised by which embryos have been

broken down into separated cells and the behavior of the dissociated cells studied *in vitro.* Cells also have been broken down into their structural components (organelles), and the chemical performances of the organelles have been observed in suitable solutions. Thus, by one means or another, organisms have been "taken apart" and analyzed.

Many fundamental answers to biological problems have come from biochemical studies. The molecular changes which take place during development have been followed in various ways. For example, the cells of an embryo may be "homogenized" by grinding them with crushed glass, or stirring them with a blender, or subjecting them to high-frequency vibrations. Next the constituent chemicals of the protoplasm can be separated by taking advantage of their physical differences such as size (dialysis and filtration), density (centrifuging), adsorptive and solubility properties (chromatography), or electric charge (electrophoresis). Finally the fractions thus obtained can be analyzed chemically. In other studies the molecules are first "labeled" with radioactive isotopes so that their history within the cell or embryo can be traced. Again proteins and other antigens have been traced by their immunological properties. In this method sensitized warm-blooded animals such as rabbits and guinea pigs are used in the assay. Molecular biology, however, is still in its infancy, and, although numerous insights have been gained, the relation of the findings to morphogenesis is not always clear. Often what is learned seems to have more to do with the physiological processes of living than with the morphogenetic processes of developing.

Much is being learned today from the study of simple organisms, such as bacteria and viruses. It now appears that all living things, whether one-celled or many-celled, whether plant or animal, are composed of similar chemicals and carry on their life processes in basically the same physical and chemical manner. That which is learned from the study of bacteria applies to man. But the problem of differentiation in embryonic development transcends the life processes of those organisms which undergo little or no differentiation.

It does not follow from the exciting advances of molecular biology that the older and more strictly descriptive aspects of biology which dealt with organisms and populations have lost their importance. Analytical biology must learn from descriptive biology what questions need to be examined; and descriptive biology can be alerted by analytical biology to look for details which had been overlooked previously. Furthermore, we must keep the wide aspects of descriptive biology in mind if we are not to lose sight of organisms as organisms, the very thing which as biologists we are concerned to understand. Let me make the point clear. When we analyze a whole into its parts, unless we are careful we lose sight of the whole and of the relation of the parts to the whole. For example, if we examine the stones of a mosaic, we lose the art of the mosaic. It does not exist at the level of the individual stones. When a phonograph record is analyzed, we discover a wavy groove. We may measure the amplitude, frequency, and wave pattern of the groove, but the music is no longer there. A book reduced to its letters is not literature. In the same manner, a many-celled organism which has been dissected to its cells, or a cell which has been analyzed to its molecules, is no longer an organism or a cell. The object which we desired to understand has vanished.

The chapters which follow are built around descriptive embryology as a core. We shall be concerned with the changes in structure which take place when an organism develops. But we shall also weave in at appropriate places some accounts of the experiments-physical, chemical, and surgical-which have been performed in trying to explain these changes. Explanations, however, are never complete. There will always be mystery. The more that science understands, the more surprising it becomes that so many complexities should exist in the universe. Especially, it is beyond our present comprehension how so many processes should work together in the production of a living creature.

Let us turn to a brief review of the history of embryology as an explanatory science.

HISTORICAL ORIENTATION

Epigenesis and Preformation

How did the embryologists of the past account for the origin of a new living being? Aristotle (384-322 B.c.), the "Father of Zoology," described the development of the chick in the egg as he and others had seen it with their naked eyes. The parts of the embryo, he said, arise in succession: first the heart; then the blood; then the blood-vessels which come from the heart; and, lastly, the various organs of the body which form around the blood vessels by processes of condensation and coagulation. Aristotle called this process of the superposition of structure upon structure during development *epigenesis.*

But how can organized complex structure arise out of simplicity? Aristotle believed that there is a sort of soul, a nonmaterial purp-osive principle, which guides development. He called it "entelechy." He identified it with the "form" of the embryo as distinguished from the matter which composed it. He supposed that the soul or form comes from the father at the time of procreation and that the mother supplies the matter or soil in which the embryo grows.

Aristotle's doctrines were accepted as authority through most of the medieval ages. They appear in the writings of Fabricius (15371619) and Harvey (1578-1657), who described development as "an epigenesis through the superaddition of parts." The de-termining cause, or soul, so he thought, is present in the blood.

With the advent of the microscope, an alternative doctrine concerning the nature of development gained ascendency. In 1673 Malpighi presented a paper before the Royal Society of London on "The Formation of the Chick in the Egg," in which he asserted that with his own eyes he had seen the chick in the unincubated egg and, hence, that development is not a process of generation, but rather one of growth. This theory, originally termed unfolding (evolution), is now referred to as the *preformation* doctrine.

The theory was soon accepted by all biologists of note and dominated biological thought for the next hundred years. Some held it on the basis of supposed observations. Thus Swammerdam, the early student of insect structure, reported that the parts of the butterfly are discernible in the chrysalis (eggs and pupae were not clearly distinguished at this time), and Buffon asserted, "I have opened a large number of eggs at different times, both before and after incubation, and I am convinced by my eyes that the chick exists in the center of the cicatricule at the moment the egg leaves the body of the hen." Others held the theory of preformation as a matter of philosophy or theoretic necessity.

It was in the spirit of the day to be logical at all costs. Thus arose the "emboitement" or encasement theory of Bonnet and Swammerdam. According to this, successive generations of individual organisms preexist one inside the other in the germ cells of the mother. It was estimated that as many as 200 million years of human beings were present, already delineated in the ovaries of Eve. Those who believed that the miniature organisms were present in the egg were known as ovists.

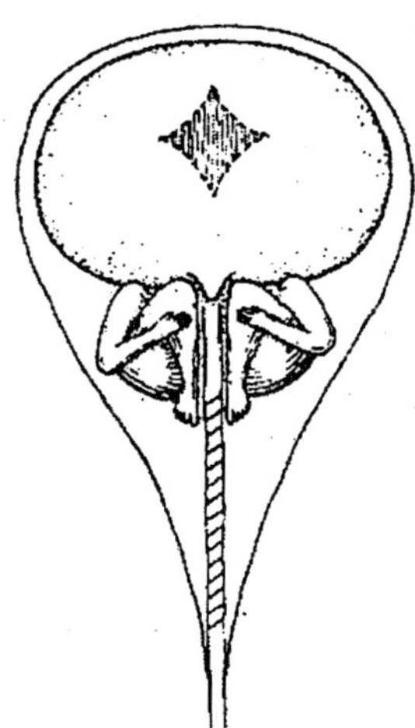

Figure 1.1 : The human sperm cell.

But two views of preformation are possible. In 1677 Leeuwenhoek (or an associate) saw sperm cells for the first time in the seminal fluid, and Hartsoeker drew a figure of what he interpreted to be a miniature manikin in the head of a sperm. The sperm was thought of as the seed (whence the name "sperm"), and the egg was the soil in which the seed was planted. Those who held this view were known as animalculists or spermists. The ensuing controversy lasted for a hundred years.

Recovery from the excesses of crude preformation began with the publication by Kaspar Wolff in 1759 of his "Theoria Generationis." First philosophically (he was moved by Leibniz's current discoveries of a mathematics of change), and then by careful observation, he demonstrated that the complete chick does not exist preformed in the unincubated egg, but that on the contrary the organs form successively in an epigenetic manner. His observations were abundantly confirmed by the great embryologists of the early nineteenth century, notably Karl Ernst von Baer (1792-1876), to whom embryology owes a great deal. Today it is clear that development is epigenetic. There is no preformed organism in either the egg or sperm.

Mosaic versus Regulative Theories of Development

During the seventies and eighties of the nineteenth century, great progress was made in the techniques of hardening, sectioning, and staining tissues and in the knowledge of the structure and functioning of cells. The complex process of cell division by mitosis was worked out by Flemming, Strasburger, and others. It soon became evident that the chromosomes of the nucleus have an important part to play in the processes of heredity and development.

In 1883 August Weismann, impressed by the remarkable precision of mitosis, proposed the theory that early development involves the orderly unpacking of an embryo already preorganized, if not actually preformed, in the chromosomes of the nucleus. He spoke of his theory as "the architecture of the germplasm." He postulated that there are units of heredity and development which he called "ids." The first cell divisions, he said, are differential; that is, they first separate the ids for the right side from those for the left, then the ids for the anterior end from the posterior, and so on for several divisions of the egg. After that, so he thought, interactions between parts and further differentiations make epigenetic development possible. Mendel's "hereditary factors"-we now call them genes-differ from Weismann's ids in that they do not segregate. Every cell during early development receives the complete set. Weismann at this time was not aware of Mendel's work.

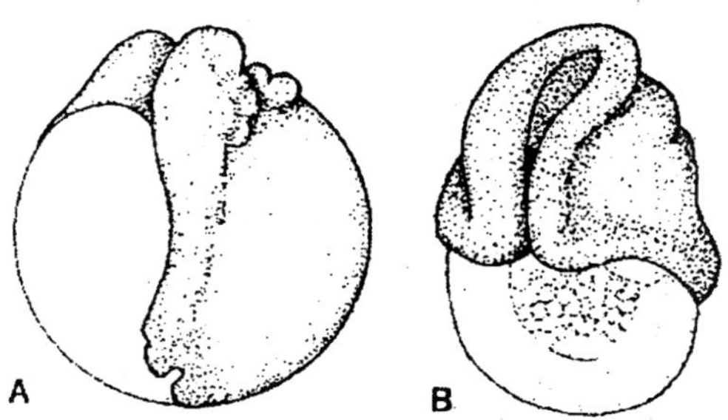

Figure 1.2 : "Half embryos" of the frog which Roux obtained in 1888 by destroying the nucleus of one cell at the two-cell stage.

In order to test Weismann's theory of the germplasm, Wilhelm Roux in 1888 performed a classic experiment which may be viewed as marking the beginning of the science of experimental embryology. He took a frog's egg at the two-cell stage of cleavage and touched one of the two cells with a hot needle thus destroying the nucleus. He observed that the uninjured cell continued dividing and developed into what he interpreted to be a one-half blastula, a one-half gastrula, and ultimately a one-half embryo. His work seemed to confirm the hypothesis that the early cell divisions of the egg are differential, a process of the orderly dispersal of the materials of the nucleus. Finally there results a ball of cells (blastula) in which each cell, so he supposed, possesses just its own potentialities for development. Roux sometimes observed a sort of reorganization of the embryo, but he dismissed this rather lightly as a secondary process of regeneration, or "post-generation," as he termed it. This general viewpoint of *Weismann and Roux, namely, that parts self*differentiate, is known as the *mosaic theory* of developmenı.

Three years after Roux's famous experiment on the frog's egg, another German scientist, Hans Driesch (1891), performed a somewhat similar experiment on sea urchin eggs. He put eggs at the two-cell stage in a vial of seawater and shook them sufficiently so that some of the cells broke apart. He then isolated the blastomeres in glass bowls and went to bed, expecting in the morning to see one-half blastulas and ultimately one-half larvae. But what was his surprise to discover that the one-half blastulas, which formed first, closed themselves into whole blastulas, and then became whole gastrulas and finally whole larvae of one-half size! He repeated the experiment with eggs at the 4-cell stage and even at the 8- and 16-cell stages. In some cases he obtained whole larvae of reduced size.

Driesch reasoned that, contrary to the theories of Weismann and Roux, the early cleavages of the egg are equal (equational), "a quantitative division of homogeneous material." Hence at an early stage the blastomeres have equal potentialities. Their fate, he said, is determined by their position in the whole. Together they constitute an *"harmonious equipotential system."* Development such as Driesch observed in sea urchin eggs came to be called regulative development, and the eggs which were capable of performing thus were called regulative eggs. To account for the guidance of the course of development, Driesch resorted, as had Aristotle long before him, to the idea of a nonmaterial purposive principle, or entelechy. A vitalistic viewpoint such as Driesch's is, of course, a matter of faith; but so is the opposite viewpoint of mechanism.

Thus began a protracted controversy in biology between those who thought that mosaic development is primary and those who considered that regulative development is the basic method by which a new individual comes into existence. The controversy still has its repercussions.

Cell-lineage Studies

The controversy drove embryologists to their laboratories and to experiments on living eggs. It was soon discovered (first by Whitman on the leech, 1878) that many eggs cleave, that is, divide into cellular units, in so stereotyped a fashion that a "cell lineage" can be traced from cell generation to cell generation. Finally, it was discovered that a particular cell gives rise to a particular tissue of the embryo or larva.

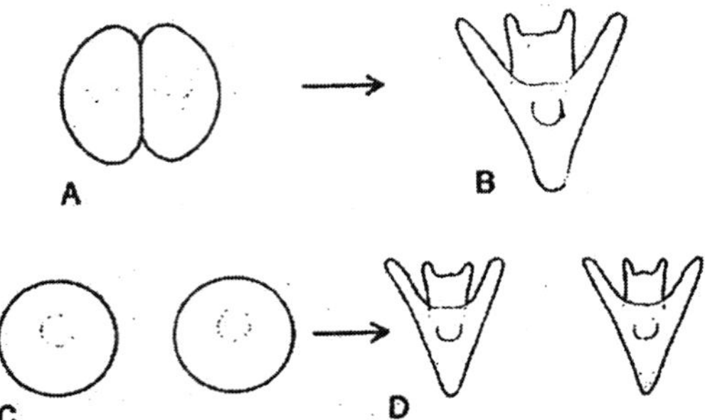

Figure 1.3 : Driesch's experiment, 1891, in which he separated the blastomeres of sea urchin eggs at the two-cell stage by shaking them in vials of seawater. A and B. Controls-whole larva from unseparated blastomeres. C and D. Experimentals-two whole larvae of half size from separated blastomeres.

In 1890 E. B. Wilson of Columbia University and E. G. Conklin, later of Princeton, were both working at Wood's Hole on Cape Cod. The former was working on the eggs of the marine annelid *Nereis,* the latter on those of the mollusk *Crepidula.* When they compared notes, they were excited to discover that these members of two entirely different phyla go through the same sequence of cell division and develop their several tissues from the same cell sources. The same symbols could be applied to designate the blastomeres. Embryology had bridged a gap between two animal phyla which the study of adult structure had not spanned.

Cell-lineage studies became the vogue, and many who later became leaders in zoology earned their Ph.D. degrees by such investigations. For a time it appeared that the regularity of cleavage must have some important and necessary relation to normal development. However, there were notable exceptions, and it seemed necessary to classify eggs into those with determinate cleavage (mosaic eggs) and those with indeterminate cleavage (regulative eggs.

Nuclear versus Cytoplasmic Theories of Development

Weismann elaborated his theories of development in terms of the nucleus, the central body which all cells possess. The early

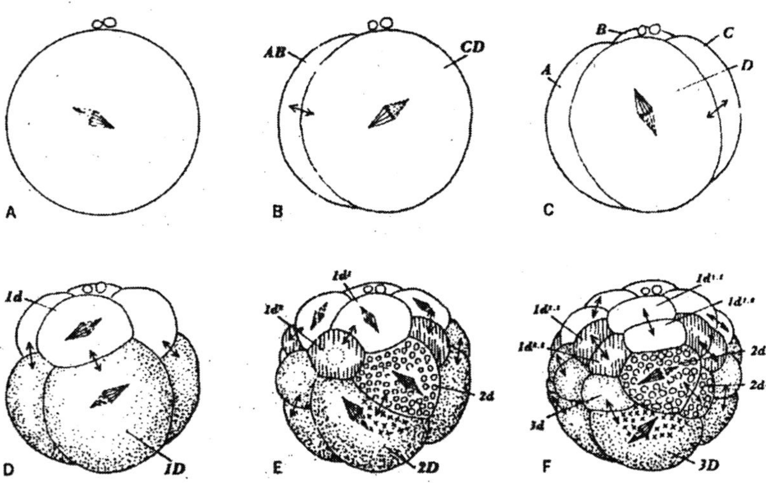

Figure 1.4 : Diagram of cell lineage in the spirally cleaving eggs of primitive annelids and mollusks. Note that the mitotic cell divisions are alternately clockwise (dextral) and counterclockwise (sinistral). The conventional labeling of individual blastomeres is shown. In E and F the fates of blastomeres are indicated by symbols: anterior ectoderm, unmarked; posterior ectoderm, finely stippled; cells of the ciliated girdle, vertical lines; cells of the ventral plate which give rise to the nervous system, marked with circles; future mesoderm, marked with crosses; endoderm of the gut, coarsely stippled.

divisions of development, he said, are differential; that is, they are such that each body cell receives only that part of the nuclear "germplasm" which determines its fate. The cell divisions by which the future germ cells increase in number, so he said, are equational, for each germ cell retains the entire germplasm of the race. Weismann included this in a principle which he called "the continuity of the germplasm." The body "which bears and nourishes the germ cells is, in a certain sense, only the outgrowth of one of them." Weismann had trouble accounting for the facts of asexual reproduction and regeneration.

It was soon shown, however, that, contrary to the nuclear views of Weismann and Roux, it is the cytoplasm, namely, the protoplasm which is outside the nucleus, and not the nucleus which undergoes differentiation during early development. Driesch in 1892 compressed sea urchin eggs between glass plates and found that the direction of the cleavage divisions could be altered so that the 8- and 16-cell stages, instead of becoming balls, became flat plates. The distribution of the nuclei to the cells of the embryo had been changed, but the distribution of the cytoplasm was unchanged. On release of the pressure, the plates of cells rounded up and developed into normal larvae. In 1896 Wilson performed similar experiments on the eggs of the annelid worm, *Nereis,* eggs with determinate cleavage, and often obtained larvae with only slight abnormalities.

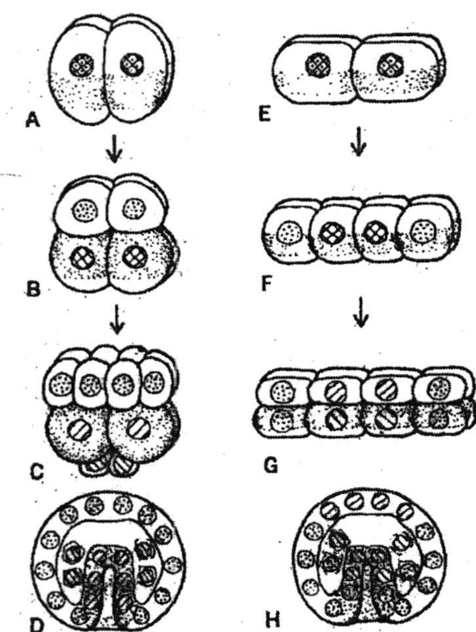

Figure 1.5 : Driesch's experiment, 1892, in which he compressed sea urchin eggs between glass plates, thus altering the distribution of the nuclei to the blastomeres. The pattern of the cytoplasm remained unchanged. A to D. Controls. E to H. Experimentals. On release of pressure, the blastomeres rounded up and developed normally.

In 1895 Morgan repeated Roux's experiment. He pricked one of the first two blastomeres of a frog's egg with a hot needle; but instead of leaving the egg undisturbed with the injured blastomere attached, he inverted the egg so that the cytoplasm of the uninjured blastomere flowed and rearranged itself. The uninjured half developed into an essentially normal whole embryo. Schultze found that if a frog's egg is inverted in the two-cell stage and held in the inverted position, a flow of cytoplasm takes place in each blastomere and a double-headed embryo may result. It must be concluded therefore, that it is the cytoplasm, and not the nucleus, which develops the pattern of the embryo. For a time it seemed that the nucleus played no part in development!

Organ-forming Stuffs

Strong support for the view that the cytoplasm, rather than the nucleus, is the basis of embryonic differentiation came from observations on the movements and final distribution of visible "stuffs". within the cytoplasm of various eggs. Boveri (1901) found that in the egg of the sea urchin, *Paracentrotus,* fertilization is

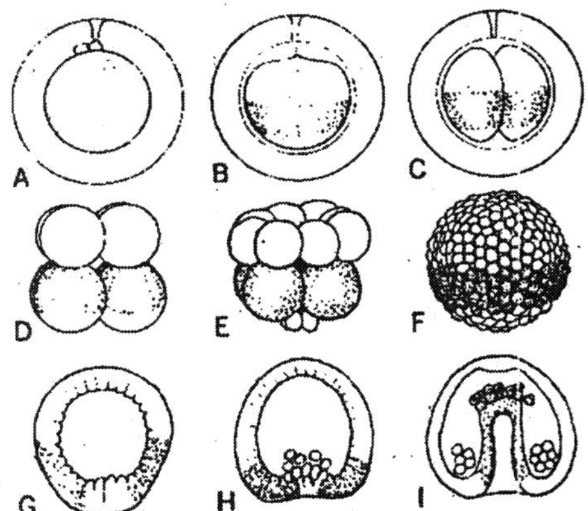

Figure 1.6 : The early development of the sea urchin Paracentrotus as described by Boveri, 1901. A. Unfertilized egg. B. Egg after fertilization. C to F. Cleavage stages. G to 1. Sections of blastula and gastrula.

followed by a flow within the cytoplasm. One result of this is that a broad superficial band of reddish-purple pigment collects just below the equator. (Most eggs have two opposite, unlike poles known as the animal and vegetal poles; the equator is midway between them.) Thus this egg is divisible into three regions: the animal hemisphere, which is pigment-free, and which becomes the outside cell layer (ectoderm) of the embryo; the subequatorial girdle with its superficial pigment, which becomes the gut and its derivatives (endoderm and mesoderm); and the small pigment-free area surrounding the vegetal pole, which gives rise to loose mesenchyme cells. These migrate within the embryo and weave the larval skeleton.

Conklin described five visibly different "organ-forming stuffs" in the eggs of the tunicate *Styela (Cynthia)*. In the unfertilized egg these are not fully segregated, but after the entrance of the sperm, an active streaming takes place. The materials become arranged in a bilateral pattern which foreshadows the bilateral symmetry of the embryo.

This process of determining the fates of regions of an egg by events which take place within the cytoplasm has been termed *localization* (the restriction of developmental capacities to particular regions) or *segregation* (the separations of regions of unlike capacity as by cell boundaries). Experiments seemed to confirm the theory, for fragments of sea urchin and other eggs which possessed each of the visibly different regions were able to develop normally, while fragments lacking one or more of the stuffs commonly produced defective larvae.

At first it was thought that the pigment itself is an organ-forming stuff (morphogenetic substance). Then in 1908 Morgan subjected the eggs of another sea urchin, *Arbacia,* to a strong centrifugal force and discovered that the pigment could be displaced from its natural position to a new location in the egg. Nevertheless, the manipulated eggs cleaved in accord with their original axiate pattern and developed quite normally, except that the pigment was now located in another part of the larva. These and similar observations on other eggs made it necessary to revise the original theory: It is not the visible stuffs (which may be wastes or by-products), but the invisible cytoplasmic substrate, or "ground substance," which is the basis of embryonic differentiation. Jacques Loeb expressed this viewpoint well when he referred to the cytoplasm of the uncleaved egg as "an embryo-inthe-rough." Recent work has emphasized the role of the outer cytoplasmic layer or *cortex* in embryonic differentiations.

Embryonic Inductions

From the time of the first experiments concerning development,

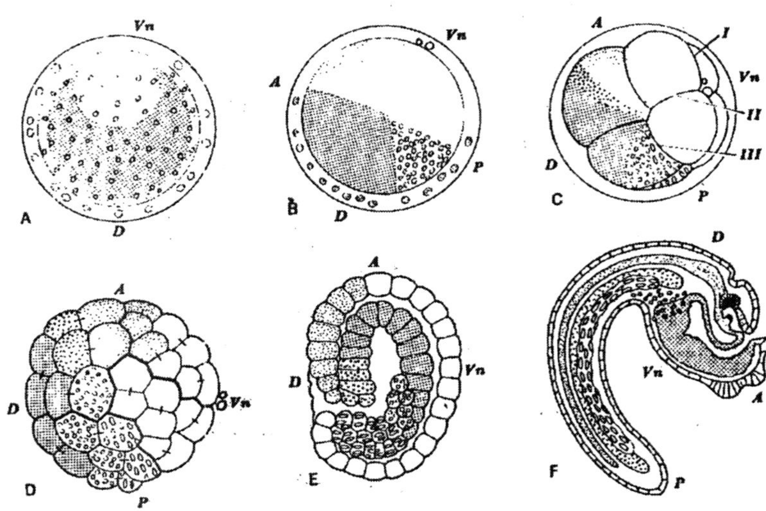

FIGURE 1-7 "Organ-forming stuffs" in the egg of the tunicate, Styela, as described by Conklin, 1905. From the right side. Vn, ventral; A, anterior; D, dorsal; P, posterior. A. Unfertilized egg. The large oocyte nucleus ("germinal vesicle") is filled with clear nuclear sap. It is close to the animal pole. Clear cytoplasm containing yellow granules (small circles) forms an outer layer surrounding a central mass of gray yolk. 6. Fertilized egg just before the first cleavage begins. Polar bodies have been given off. A rearrangement of "stuffs" has taken place, including a gathering of the yellow yolk granules to form a yellow crescent on the future posterior side of the egg. C. Eightcell stage. The first cleavage plane (I) divides the egg into symmetric right and left blastomeres. The second cleavage plane (11) divides anterior blastomeres from posterior blastomeres. The third cleavage plane (111) separates a ventral quartet of blastomeres around the animal pole from a dorsal quartet surrounding the vegetal pole. At this stage six regions can be distinguished. These are represented schematically thus: Unshaded-clear plasma, future epidermis; lightly stippled-bright gray plasma, prospective nerve cord; coarsely stippled-less brilliant gray, notochord; darkly stippled—slate gray yolk, endoderm; small circles-bright yellow, prospective mesenchyme; ovals-dark yellow, future muscle cells. D. Sixty-four-cell stage. The heavy lines mark the second and third cleavage planes. The short transverse lines connect daughter cells of the sixth cleavage. A slight flattening on the dorsal side indicates that gastrulation is about to begin. E. Optical section through a gastrula. Future neural and epidermal cells remain on the outside (ectoderm). Endoderm, notochord, mesenchyme, and muscle are inside. F. Late embryo showing. the fates of the several formative "stuffs."

it has been evident that there are interactions between the parts of the developing embryo. Interactions assure that the parts will develop in their proper positions. Interactions also regulate their relative sizes. This is true even when the parts have been experimentally disarranged.

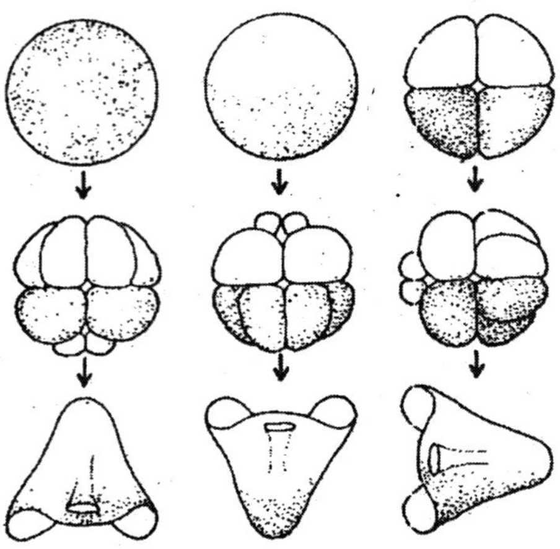

Figure 1.8 : Morgan and Spooner, 1909, centrifuged uncleaved eggs of the sea urchin Arbacia. The heavy pigment granules (shown by stippling) were driven toward the centrifugal pole. Except for this dislocation of pigment, development continued normally.

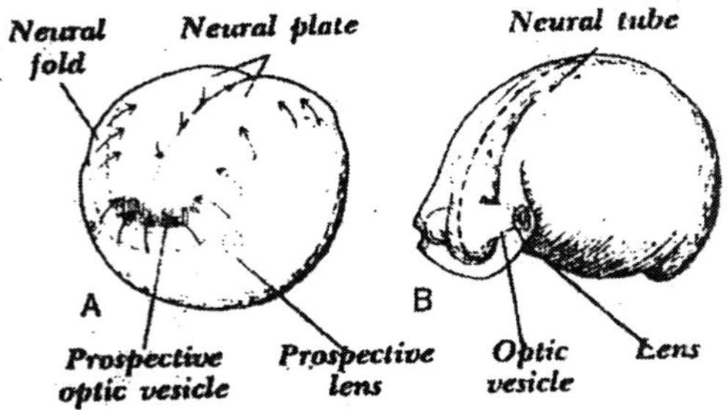

Figure 1.9 : The development of the eye in amphibians. A. Open neural plate stage. The location of the prospective optic vesicles (future retinas) is shown by vertical lines, that of the prospective lenses by stippling. B. As a result of the folding of the neural plate to form the neural tube, the bulging optic vesicles come into contact with the epidermis and induce the production of lenses.

A classic example of interaction, or at least of action and reaction, was the discovery by Spemann in 1901 of the development of the lens of the vertebrate eye. The retina and lens come from separate embryonic sources: the retina from the sides of the forebrain, which in turn comes from the neural plate; the lens from head epidermis. When the neural plate folds to form the brain, the epidermis comes into contact with the bulges of brain tissue which are destined to form the optic cups (future retinas). What would happen if these two tissues should fail to come into contact with each other? Spemann and others studied this problem in amphibians by surgical procedures. They found that the epidermis of the future lens region possesses a competence to form a lens; but if no contact is made, a normal lens does not usually form. The optic cup is said to induce or evoke the epidermis to form a lens.

The area which possesses the competence to form a lens is not sharply defined at first. Indeed, in some species epidermis from as far away as the sides of the trunk has some capacity to respond to the stimulus of the optic cup and form a lens. This has been shown by grafting an optic cup beneath flank epidermis and also by grafting flank epidermis to the head in place of head epidermis. Thus the competence of the epidermis and the inductive stimulus arising in the optic cup constitute a reaction system which assures that the lens will form just where it should, and that a unified eye will result. In the chapters which follow, we shall note many other examples of embryonic induction.

Organismic Viewpoints

The goal of experimental embryology has always been to give a causomechanical account of the processes of development. The earlier theories were predeterministic. Although they did not always postulate a "little man" in each germ cell, as did the theories of the early microscopists, yet they proposed an "architecture of the germplasm," or a precise sequence and pattern of cell divisions, or a cytoplasm that is an "embryo-in-the-rough." Even Driesch's nonmaterial, vitalistic entelechy was a purposive, predetermining principle. But all these older theories failed to come to grips with the first and basic problem of development; namely, whence comes organization in the first place? How does a new individual get its start? The trouble with the older theories was that they did not begin at the beginning. They begged the basic question of development by assuming that the embryo is represented in some manner in the germ. Some students of heredity still seem to talk this way, but as we soon shall see, heredity is a very different thing from predetermination.

A completely new insight into this most fundamental of all embryological problems came indirectly through studies on the regeneration of lower animals. Regeneration has always proved a stumbling block to biological theorizing, especially to theories of the mechanistic type. If a part can form a whole-almost any part in the case of some of the lower organisms-then development must be completely epigenetic; the new individual cannot be predetermined in the part. And here vitalistic assumptions do not help us at all, for they are also doctrines of predetermination.

Roux (1885) tried strenuously to give a causomechanical account of development on an epigenetic basis, but he gave it up in favor of his mosaic theory. Morgan struggled with the problems of regeneration for years before he turned his attention to genetics in the hope that genes would supply the clue. Loeb clearly recognized that regeneration was the crucial problem in his attempt to give a mechanistic account of life.

A major breakthrough came as a result of the work of H. V. Wilson and C. M. Child. Wilson (1907) found that when sponges are pressed through bolting cloth into a dish of seawater, the individual cells become dissociated from one another and settle to the bottom. There they adhere together in clumps and shortly reorganize new and unified sponges. Similarly Child found that when he fragmented a large hydroid polyp *(Corymorpha)* by grinding it lightly in seawater with sharp ground glass, the isolated cell's which still remained alive settled to the bottom, and there aggregated and reconstituted new individuals. Here, beyond equivocation, were heaps of cells establishing a new pattern of wholeness. This is, indeed, epigenesis from a state near chaos. But of course it is not chaos, for the cells which are capable of such regeneration are organized living things which contain the capacities of their species. Moreover, there are different kinds of cells; and each cell, when it comes into contact with another cell, has the tendency to adhere to it and then to sort out with other cells of its own kind. Facts of this sort help explain how, out of a disordered mass, a new individual takes form.

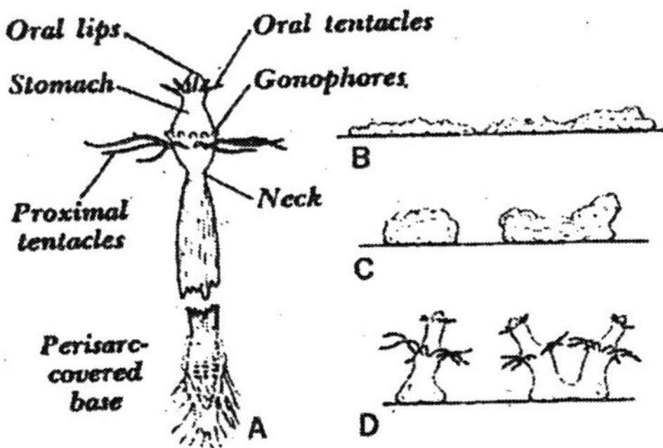

Figure 1.10 : Reconstitution of a hydroid polyp as described by Child, 1928. A. Polyp of Corymorpha showing zones of the hydranth. B. Dissociated cells adhere. C. They draw together into clusters. D. They differentiate into new polyps.

On the basis of numerous experiments of many kinds, Child developed his theory of "axial gradients." He proposed that the first step in regeneration, as in all development, is the establishment within the developing cell, or group of cells, of some sort of quantitative differential from one region to the next. Between adjacent levels of the gradient, so he supposed, there is a relationship of "physiological dominance" and "physiological subordination." During regeneration (or development) the dominant region, having become free from the control of adjacent regions, does what is its nature to do. In the case of the hydroid polyp it becomes oral lips. At the same time, under the influence of the dominant region, adjacent regions become in sequence other things. In the case of the marine polyp, *Corymorpha,* the subordinate regions produce in succession oral tentacles, the stomach, basal tentacles, the neck, etc.

Reproduction, then, in Child's view, starts with isolation-actual isolation in the case of development from fragments or eggs, but "physiological isolation" in the case of a budding polyp. That is to say, a region buds and develops a new whole when it escapes from the physiological dominance of its parent's body. The principal criticism of Child's theories has been that, in spite of all the evidence he accumulated concerning the many kinds of gradients in many organisms, he failed to discover a basic gradient of which all other gradients are but effects and expressions. But even if a basic physical or chemical gradient were to be identified-and today we may not be far from doing so-this in itself would not explain dominance and subordination. Such words do not explain anything. They serve to pinpoint what needs to be explained.

A concept which is similar in some ways to axial gradients is that of an embryonic field. This is, after all, just a word to remind us that patterns of a quantitative and geometric nature exist within the developing embryo, and that, in relation to these patterns, differentiations take place in an orderly manner. A field is not a mosaic work in which each part possesses its own independent capacity for selfdifferentiation. Rather, it is a whole in which each part differentiates according to its spatial (and temporal) relations to other parts. A field is something more than embryonic induction, or an interaction between parts, for it precedes the parts, and it is in relation to the field that the parts become different. Some prefer to speak of a field as a "pre-pattern." This adds nothing to our understanding.

Organismic views such as these are truly epigenetic. They do not assume a pattern or organization to begin with. Moreover, they are not deterministic, as are mechanistic and vitalistic hypotheses. They allow the regenerating part, or the developing embryo, to find its own way as it goes. It may develop as one whole; but if there is some interference with the gradient, or some distortion of the field, it may equally well give rise to an abnormal creature, or to a duplication of parts, or to twins.

The Nature of Differentiation

What is embryonic differentiation? It is, of course, the process by which a region or cell of a developing egg or embryo becomes demonstrably different from other regions or cells. It becomes a *rudiment* or *primordium* (German: *Anlage).* These terms are synonyms. But before visible differentiation has begun, some invisible process or event must have taken place.

In the older literature of experimental embryology this invisible process was described in negative and conceptual terms as a "segregation" or "localization" of the potentialities of development. Each region was thought of as being at first "totipotent," that is, as having the capacity of becoming any part of the embryo. Then, it was supposed that the potentialities became more and more restricted until finally no potentiality remained, except the potentiality to become the one thing which it was the region's fate to become.

This line of reasoning, however logical it may be, has proved sterile. Today, differentiation is described in positive and concrete terms as a straightforward coming-into-being of molecules which did not exist at first. Generally speaking, these molecules are macromolecules (nucleic acids, proteins, etc.). Differentiation, then, is the controlled process by which regions or cells of like hereditary capacities become different in their biochemical composition. The macromolecules include enzymes at the surfaces of which the chemical reactions of the cell take place.

The result of differentiation is *determination,* the decision as to what a given region will become. Determination is a progressive process. At first, it is "labile" and reversible; but with the passing of time it becomes increasingly stable and irreversible. Finally, the region or cell, having become fully determined, gives rise to the structure which it has acquired the power to become.

The Role of Heredity

Experiments on embryos have thus established that (1) development is epigenetic; (2) although it is regulated by the nucleus, it takes place primarily in the cytoplasm; (3) it involves interaction between parts; (4) the parts arise within gradient patterns or "fields"; (5) differentiation is in essence the development of the macromolecular pattern (including the distribution of enzymes) within the cell. (Much of the differentiation takes place before cleavage in mosaic eggs but after cleavage in so-called regulative eggs.) However, a basic question still eludes us: What supplies the *possibilities* of development? Why does a human egg, given a normal environment and opportunity, become a man and not a mouse? The answer, of course, is *heredity*, that precious something which is passed on to the offspring by its parents.

What is heredity? One of the triumphs of modern science is the discovery that what is passed on by heredity consists of an assortment of huge molecules-macromoleculeswhich are contained notably, but not exclusively, in the nucleus of every cell, and which are transmitted from one cell generation to the next. These essential macromolecules include proteins which, in their role as enzymes, control the chemical reactions of the cell. They also include nucleic acids. Some nucleic acids (DNA) are the genes which serve as patterns or templates for their own replication and for the production of other nucleic acids (RNA). The latter engage in the manufacture of the chains of amino acids (polypeptide chains) from which the proteins are formed. The nucleic acids pass on "information" in the form of an alphabetic "code" of four nucleotides. The nucleotides are arranged in a linear fashion like beads on a string. The sequence in which these nucleotides follow one another is significant, for the sequence of the nucleotides in the nucleic acids determines the sequence of the amino acids which make up the polypeptide chain.

Note that the modern understanding of heredity is not a doctrine of preformation. It is not even a theory of predetermination, for an egg is subject to accidents and to the influences of its environment as it develops. What the egg is to become is not settled in advance. Heredity supplies only the *possibilities* of what it may become. Hereditary materials are blueprints, as it were, not of the individual that is to be, but of the developmental process which, under the influence of the circumstances which surround it, will bring the new individual into being.

In our enthusiasm for our new insights into the chemical basis of heredity, we must not lose sight of the fact that the organism as a whole nurtures and protects hereditary material. It is the entire organism which survives in the "struggle for existence." It is not the organism in any static sense which carries on. Rather, it is the entire *life cycle* as a system in both time and space-egg, embryo, larva, juvenile, adult-that endures. Every stage must be adapted to survive, or the race will come to an end. The proteins and nucleic acids are the thread of continuity which runs through the life cycle. These are the materials which are transmitted from one generation to the next.

Every life cycle is the product of a long evolutionary history and can be understood only when its past is taken into account. Just as the art of a mosaic, the music of a phonograph record, or the literature of a book is an achievement, so every life cycle is an achievement. It is the product of the creativity of nature, as truly as a painting or a symphony is the product of the creativity of a human mind. This is true whether we account for creativity in terms of time and chance (that is, by mutation and selection) or by the hypothesis of a Creator. The production of even the simplest life cycle has required untold millions of years. The macromolecules which are peculiarly of biological importance are those which have contributed to the continuation of living things on earth. They are "vital," not in the sense that they possess some nonmaterial principle or entelechy, but because they are the blueprints of successful living. As such they possess a significance which purely chemical systems do not possess.

Herein lies the distinction which sets biology apart from the physical sciences, and which we are apt to forget when we analyze an organism in terms of the elements which compose it and the physical processes by which it lives. The characteristics of chemical molecules such as salt, sugar, or a plastic can be predicted on the basis of the properties and arrangement of the atoms which compose them. But nucleic acids and proteins have properties (based on the sequence of nucleotides or amino acids) which can be accounted for only in terms of the long history of life on earth. Indeed, such words as "history," "heredity," "information," "function," and "survival" have no real meaning in physics and chemistry. They apply to selfreproducing systems which have had a past, and which, if all goes well, may be expected to have a future. Apart from a particular living system, a specific protein or nucleic acid would not even exist; much less would it have a "reason for being." These are matters which tend to be overlooked in discussions of the physics and chemistry of life.

How far embryology will ultimately go toward providing an account of the processes of development, only time will show. There is a sense in which it cannot be expected to go all the way, for the final outcome of development is a conscious, striving being. We know that this is true, each of us with respect to himself. We infer that it is true also of other men and, in varying degrees, of other organisms. But these are propositions which transcend science.

2

Reproduction

THE MALE REPRODUCTIVE ORGANS·

The primary male organs consist of paired glands called the testes, and these produce the male gametes or spermatozoa. The testes develop in the abdominal cavity adjacent to the adrenal glands but in most mammals they migrate downwards through the body cavity into a special fold of skin called the scrotum, or scrotal

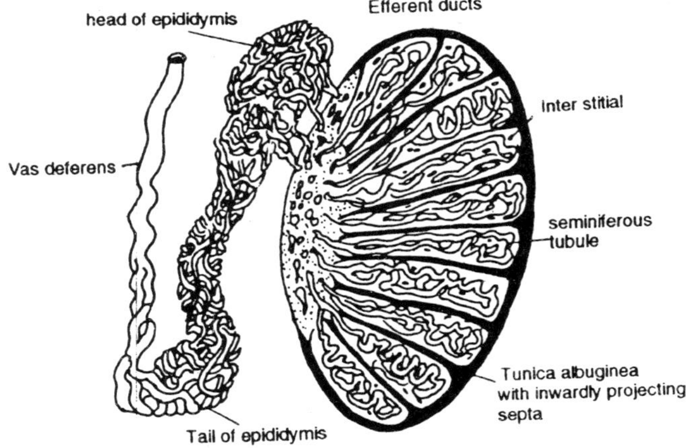

Figure 2.1 : Diagram of a vertical section of the testis.

sacs. The reason for this is that spermatogenesis will not take place at body temperature, and within the thin walled scrotum the temperature is several degrees below that of the body cavity. If for some reason the descent does not occur then the animal is infertile. In amphibia, reptiles and in some mammals e.g. the elephant, the testes remain within the abdominal cavity. The testis is guided in its descent by a long cord, the gubernaculum, which extends from the lower pole of the testis to the scrotum, and the gubernaculum anchors the mature testis in the scrotum.

The secondary sexual organs include a variety of ducts and glands which convey the spermataozoa in special secretions to the exterior of the body, so that they can deposited within the female; the secondary sex organs also, include those characters of the male which distinguish it from the female e.g. the long mane of the male' lion. These secondary sex organs are developed and maintained by,means of sex hormones produced within the testis itself. Male sex hormones are called androgens.

Structure of the Testis

The testis is a tubular gland surrounded by a fibrous capsule the tunica albuginea. It is divided into several hundred compa-rtments by means of fibrous tissue septa and each compartment contains several tubules, called seminiferous tubules. Each tubule is about 50 cms. long in man and is coiled upon itself, hence the name convoluted seminiferous tubules. All the tubules drain into one border of the testis, into larger collecting tubules which are coiled together in a mass called the epididymis which is applied to the surface of the testis. Between the seminiferous tubules inside the testis, there is connective tissue containing blood vessels and the glandular cells which are called interstitial cells or Leydig's cells and are responsible for the production of the male sex hormone.

Structure of the Tubule

In the adult the seminiferous tubule consists of a basement membrane lined by the seminiferous epithelium; this seminiferous epithelium consists of two types of cell. First the germ cells themselves, secondly the cells of Sertoli which support and nourish the germ cells. The youngest germ cells are those lying close to the wall of the tubule and these divide to produce cells which pass nearer to the lumen of the tubule where the mature spermatozoa occur. The details of this transformation are described in more detail. The Sertoli cells are slender pillar like cells attached at their base to the basement membrane. At from the lumen of the seminiferous tubules into the larger collecting tubules of the epididymis, whose walls bear a ciliated epithelium which moves the spermatozoa into the vas deferens.

Secondary Sex Organs

The spermatozoa are conveyed by the vasa deferentia to the base of the bladder where they can pass into the urethra. In structure the vas deferens is a hollow muscular tube, which, by

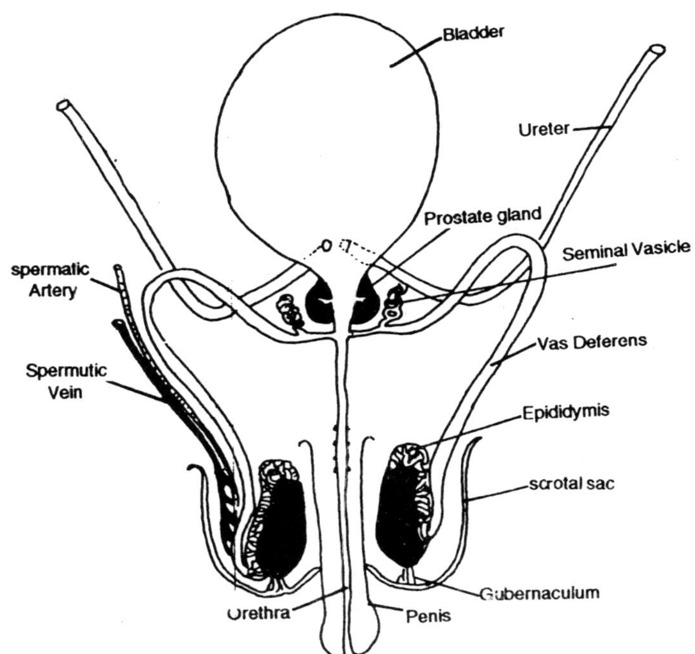

Figure 2.2 : Diagram of the male reproductive organs.

means of muscular. contraction, rapidly transports the spermatozoa before they are deposited in the female. Opening into the vas deferens before it joins the bladder neck is the duct of the seminal vesicle; the seminal, vesicles were once thought to store sperm but their function is to produce a thick secretion to provide the bulk of the fluid in which the sperms are transported. Surrounding the base of the bladder where the vasa deferentia open into the urethra is another gland, the prostate gland, whose secretions are discharged into the urethra together with the spermatozoa and the secretions from the seminal vesicles. The prostate gland produces a thin alkaline secretion; this helps to neutralize any acid urine remaining in the urethra and also to neutralize some of the acid secretions of the female vagina after the sperms have been placed into the female.

These various secretions, spermatozoa and the products of the seminal vesicles and prostate gland are together called semen; the semen is discharged through the urethra which passes through the penis. The penis, through which both urine and semen pass, contains in its walls, sponge-like systems of blood spaces which can become filled with blood so making the penis a more rigid organ so that the semen can be deposited within the female.

THE FEMALE REPRODUCTIVE ORGANS

The primary sex organs in the female consist of paired ovaries situated within the abdominal cavity. The germ cells are liberated from the surface of the ovary into the peritoneal, cavity from whence they pass into the secondary sex organs- the Fallopian tubes, the uterus and vagina, leading to the exterior of the body. Paired secretory organs, the mammary glands, are included in the secondary sexual organs of the female, and serve to nourish the newly-born mammal.

Structure of the Ovary

The ovary is attached to the wall of the body cavity by a fold of peritoneum. The free surface of the ovary bulges into the peritoneal cavity into which the germ cells are liberated. The ovary is studded with follicles in various stages of development, containing the germ cells. When the follicles are ripe they come to the surface of the ovary where they rupture.

The free surface of the ovary is covered by a thin layer of germinal epithelium from which the germ cells arise in the embryonic period. In some species of mammal it seems that even in the adult the germinal epithelium can give rise to successive crops of new germ cells which pass inwards to mature in the tissues of the ovary. The timing of the successive crops of new germ cells coincides with the rupture of mature Graafian follicles in which follicular fluid rich in the hormone oestradiol pours over the surface of the ovary. This hormone has been called a 'mitogenic' hormone because of its effect on the germinal epithelium in stimulating cell division and the formation of new crops of germ cells. Beneath the germinal epithelium is a layer of dense connective tissue, the tunica abhiginea. Beneath the tunics albuginea the thicker outer part of the ovary, or cortex, contains the follicles in various stages of development. The central part of the ovary or medulla contains a loose connective tissue containing masses of blood vessels.

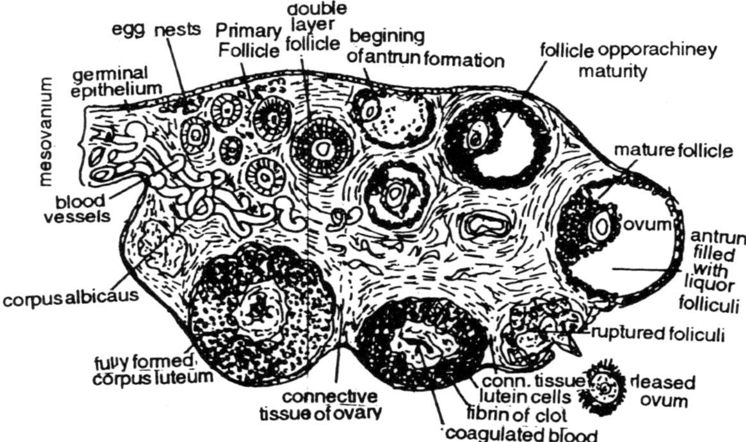

Figure 2.3 : Mammalian Ovary

The interstitial connective tissue of the ovarian cortex consists of connective tissue fibres and various types of cells. Some of these cells, large polyhedral 'epithelioid' cells are given the name interstitial cells. The number of these cells varies throughout the life of the female mammal. In some mammals with large litters e.g. rodents, there may be enormous numbers of these cells. They may arise from the walls of degenerating Graafian follicles.

Not all of the follicles in the ovary undergo the course of the development into mature Graafian follicles as described. Very many of them undergo a degenerative change called atresia in which there is a hypertrophy of the cells forming the wall of the follicle together with a degeneration of the ovum. The interstitial cells and the cells of the atretic follicles are considered to have an endocrine function.

The oviduct or Fallopian tube

The oviducts are muscular tubes which serve to convey the germ cells from the ovaries to the uterus. The outer end of the tube, nearest to the ovary is expanded, and its edge is split up into fringes, the fibriae, which are closely applied to the surface of the ovary. The lumen of the oviduct is lined by a secretory mucous membrane, in which there are many ciliated epithelial cells. The germ cells are conveyed down the tube to the uterus by means of peristaltic movements of the tube itself and by the effect of the ciliated epithelium.

The Uterus

The uterus is a thick-walled muscular structure within which the embryo develops. Its wall has three layers, an outer serous coat, a thick middle coat consisting of interlaced smooth muscle fibres (the myometrium) and an inner vascular mucous layer, the endometrium.

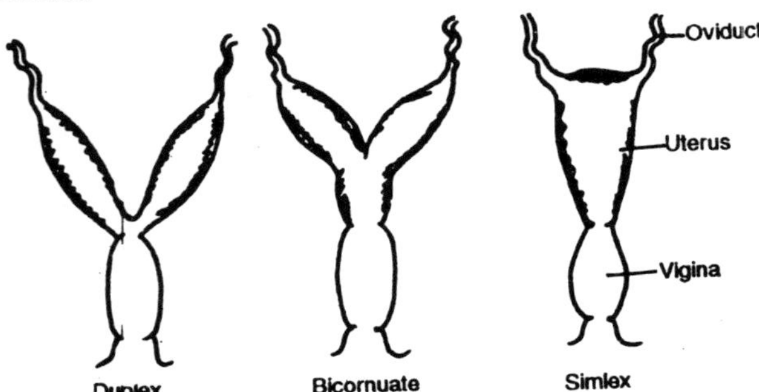

Figure 2.4 : Types of mammalian uteri.

In primitive mammals there are two uteri, each opening into the vagina and this is called the duplex condition and is found in marsupials, many rodents (e.g. rats, mice, rabbit) and bats. In most mammals the distal end of the two uteri is fused to give a biconuate uterus. In higher primates, including man, the two uteri are completely fused together to give a single organ, the uterus simplex.

The vagina is a distensible tube lined by squamous epithelial cells which connects the uterus to the outside world.

GAMETOGENESIS

The process of formation of gametes is called gametogenesis; the formation of eggs is called oogenesis and the formation of sperms is called spermatogenesis. Gametogenesis may be conveniently divided into three stages. The first stage is one in which the cells of the germinal epithelium divide and is called the stage of multiplication. The second stage is one of growth when each of the tiny cells produced in the multiplication stage grows to a larger size. The cell at the end of this stage is called the primary oocyte in the case of the female, and the primary spermatocyte in the case of the male. The third stage in gametogenesis is a period of maturation; during this period very important changes occur in the nucleus with the result that the chromosome number is reduced from the diploid to the haploid number. This reduction in chromosome number takes place in the so-called reduction division of a specialkind of cell division called meiosis. When the primary oocyte divides by the reduction division it does so unequally; the nucleus divides into two equal parts, but almost all of the cytoplasm goes with one half of the nucleus, whereas the remaining half of the nucleus has very little cytoplasm. The latter is called the first polar body. The nucleus with most of the cytoplasm is now called the secondary oocyte and it contains only the haploid-number of chromosomes. In mammals it is at this stage that the female gamete is released from the ovary; and before this gamete can be considered as fully mature another division of the nucleus has to take place, and this again is an unequal division resulting in the production of a second polar body. The production of this second polar body takes place, in a mammal, when the egg is fertilized by the sperm. We have seen that the development of the female gamete involves three stages, the first of multiplication, the second of growth ending in the formation of the primary oocyte, and the third of maturation involving the production of the secondary oocyte with polar bodies.

The first stage of gametogenesis is complete in the female embryo by the end of intra-uterine life, and she is born with all' the oogonia already formed within the ovary. The second stage of growth of the oogonia continues throughout the life of the mammal and we will now look at this growth phase in more detail.

Development of the Graafian follicle

In the cortex of the ovary of the mature mammal are many small clusters of cells called the primary follicies, which have been formed during the embryonic period from invaginations of the germinal epithelium, called sex cords. . The primary follicles are very small and there are about 400,000 in the human female at maturity. At birth the first phase of oogenesis, the phase of multiplication has already started producing small collections of germinal cells, called primary follicles. One of the cells in the primary follicle is larger than the rest and is the oogonium, whilst the smaller surrounding cells are called follicular cells. In the second phase of oogenesis, the growth phase, the primary follicle develops and changes occur, in the oogonium as it becomes the primary oocyte, in the follicular cells and also in the connective tissue which surrounds the follicles. The oogonium enlarges, its nucleus gets bigger, and a few yolk granules begin to appear in its cytoplasm. At this stage a well defined shining layer appears around the surface of the oogonium called the zona pellucida. In the primary follicle the oogonium was surrounded by a simple columnar epithelium but as the follicle grows the follicular cells multiply to produce an epithelium which is several layers in thickness. The cells of this epithelium secrete a follicular fluid which accumulates in spaces which begin to appear between the cells.

The follicle by this stage in the human female is about 2 m. in diameter and is now called the Graafian follicle. The follicle increases in size with the accumulation of more fluid within it and the oogonium is pushed to one side of the follicle where it is attached to the wall of the follicle in a group of columnar cells called the discus proligerus. The cavity of the follicle is lined by a few layers of columnar cells called the membrana granulosa. The connective tissue surrounding the Graafian follicle has become organized into a membrane called the theca (consisting of two layers, the *theca* interna and externa). Eventually the Graafian follicle may reach a size of 10 mm. in diameter in the human female and bulges from the surfaces of the ovary; by this time the oogonium has grown to its full extent and is called the primary oocyte. The fluid within the follicle is formed at a faster rate than the follicle wall grows and the follicle eventually ruptures.

By the time the follicle has ruptured the primary oocyte has undergone the meiotic division producing the first polar body and is now called the secondary oocyte. When the secondary oocyte is released it is surrounded by a few columnar cells which form the corona radiata, which may have a nutritive function similar to that of Sertoli cells in the male. When the Graafian follicle has ruptured and liberated the oocyte it collapses and the hole left by the departing oocyte becomes plugged with a blood dot. There is now a multiplication of the remaining cells of the follicle, the granulosa and theca cells. The cells enlarge and develop deposits of a yellow pigment called lutein.

The whole structure so produced is called the corpus luteum, a solid ball of-yellow pigment cells, which produces hormones which prepare the uterus to receive the fertilized oocyte.

Spermatogenesis, the formation of spermatozoa. Unlike the female, where the germinal epithelium forms the outermost layer of the ovary, in the male the germinal epithelium lines the walls of the seminiferous tubules. The cells nearest to the wall of the tubule, the spermatogonia, are the most primitive, undifferentiated cells of the tubule and they give rise to the other cells by mitotic division. As in oogenesis there are three stages in spermatogenesis.

The first stage of spermatogenesis is that in which the spermatogonia multiply by mitotic division. Some of the products of these divisions pass inwards nearer the lumen of the tubule where they enter the second phase of spermatogenesis, the growth phase. During this phase the spermatogonia become larger, producing the

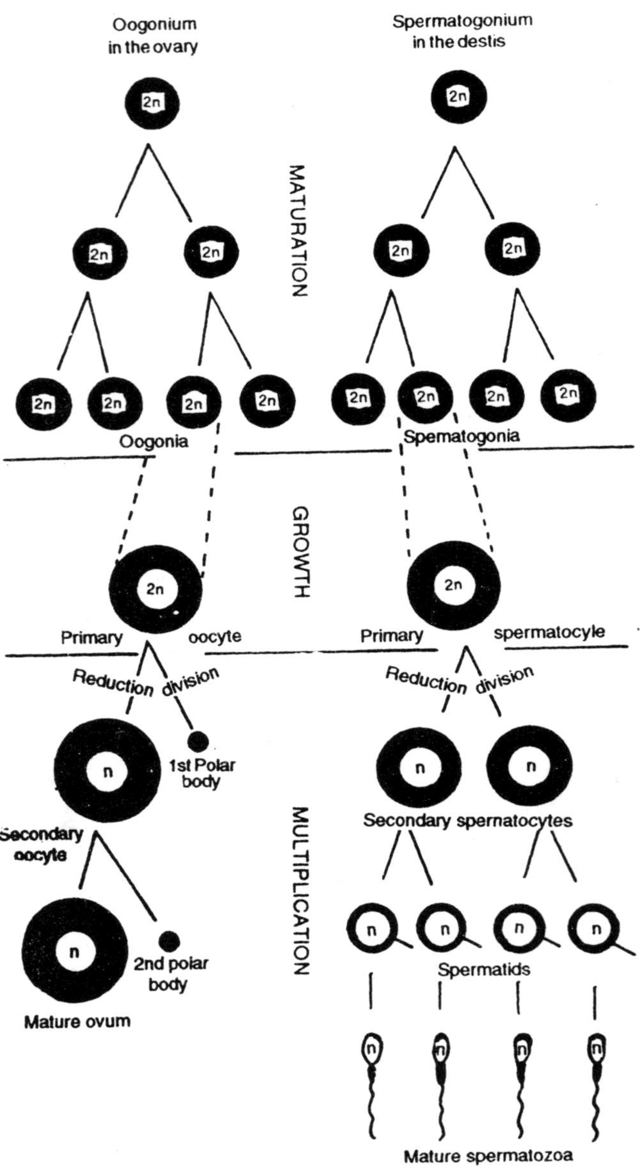

Figure 2.5 : Structure of a mature sperm.

the primary spermatocytes. In the third phase each primary spermatocyte undergoes a reduction division producing two secondary spermatocytes containing the haploid number of chromosomes. Each secondary spermatocyte then divides to produce two spermatids. The spermatids do not undergo divis on but a series of changes occur which transform the spermatid into the mature sperm. During this maturation of the spermatids they are attached to the cells of Sertoli.

The Mature Sperm (human). The mature sperm consists of a head, middle piece and tail. The head consists of the condensed nucleus of the spermatid and is a flattened ovoid structure about 5 microns long. The head is capped by a sheath of material called the head cap. In the middle piece of the sperm there is a centriole from which arises a long axial filament which passes through the middle piece and the tail. Surrounding the axial filament in the middle piece is wound a sheath of mitochondrial material, mitochondrial sheath, which is probably concerned in the respiration of the sperm. In the tail the axial filament is covered by a sheath.

The spermatozoa remain inactive until they pass from the testis. During their passage from the testis to the penis they are activated by the secretions of the accessory glands. The sperm is capable then of active swimming during which S-shaped waves pass along the tail. The energy for this is derived from the anaerobic breakdown of fructose which is present in the prostate secretions.

THE PHYSIOLOGY OF REPRODUCTION

Hormones and Reproduction in the Male

In the sexually immature mammal the primary and secondary sex organs are small and undeveloped. The growth and development of the sex organs is dependents upon the activity of the pituitary gland. The pituitary exerts its effect by the production of hormones called gonadotrophic hormones because of their growth effects upon the gonads. The gonadotrophic hormones are complex protein substances which have been isolated in a relatively pure state; their chemical structure is not yet elucidated. There are almost certainly two gonadotrophic hormone. One of these hormones is called the follicle stimulating hormone (or F.S.H.) because of its effect in the female in stimulating the growth of the follicles in the ovary. F.S.H. stimulates the growth of the seminiferous tubules of the testis and stimulates the activity of the germinal

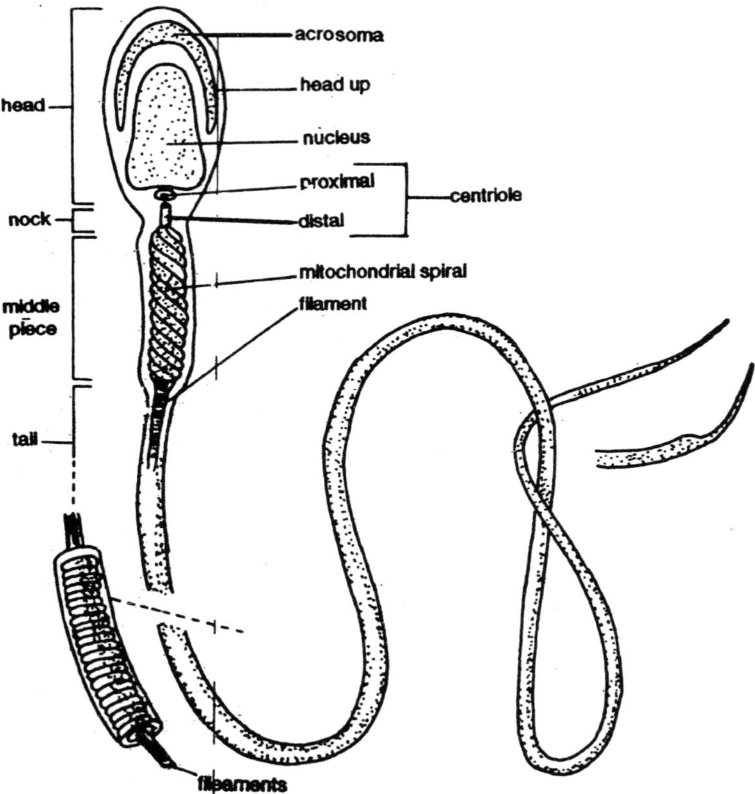

Figure 2.6 : Formula of testosterone.

epithelium. The other pituitary hormone is called the luteinizing hormone (L.H.) or interstitial cell stimulating hormone. L.H. stimulates the growth and secretory activity of the interstitial or Leydig cells of the testis. These cells produce certain steroid hormones called sex hormones, because of their effect on the sex organs.

The most important sex hormone in the male is called testosterone. However, there are also female sex hormones or oestrogens produced in the male. The effect of testosterone is to promote the growth of the various sex organs the vas deferens, seminal vesicles, prostate gland, penis etc. It also promotes the development of those other secondary sex characters which vary from one mammalian

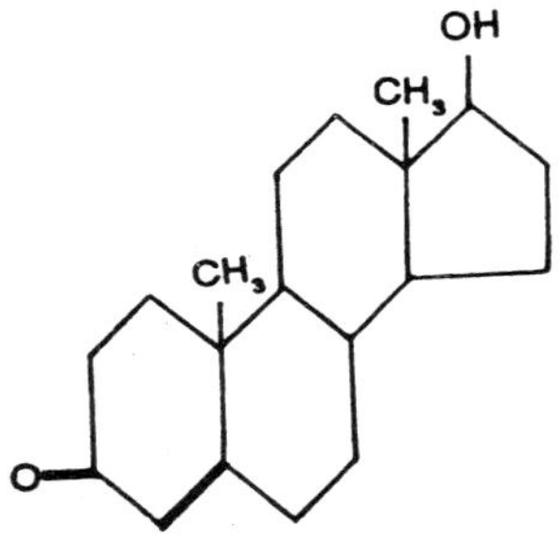

Figure 2.7 : Formula of testosterone.

species to another; in man these include the growth of the beard, enlargement of the larynx with the development of the deeper voice, the male distribution of hair on the body, and the greater development of muscle. The effect on muscle growth is due to the fact that testosterone is what is called a protein anabolic hormone, that is it promotes the retention and incorporation of protein in the tissues.

In the immature animal testosterone will promote the precocious development of the sex organs, and in those species of mammals in which the testes do riot descend into the scrotum until sexual maturity it stimulates the descent of the testis. In animals which have been castrated the sex organs gradually atrophy and the administration of testosterone can reverse these charges.

Testosterone has also a direct effect on the pituitary gland and a certain level of testosterone in the blood will inhibit the pituitary from producing gonadotrophic hormones; when the pituitary production of leuteinizing hormone falls then the Leydig cells of the testis stop producing testosterone and the blood level of testosterone falls. With the falling producing of testosterone by the Leydig cells the pituitary gland is now released from the inhibitory effect of testosterone and it begins to produce the gonadotrophic hormones again. By this feedback mechanism. the secretion of testosterone is controlled.

This effect of testosterone upon the pituitary gland also explains the varying results that experimenters have had in the administration of testosterone to animals. Some workers have found that in some animals testosterone will cause a stimulation of the germinal epithelium of the testis and the production of spermatozoa; because of the appearance of large numbers of dividing cells in the germinal epithelium the hormone has been called 'mitogenic', that is one which stimulates mitotic division within the cells. In large doses however testosterone can depress the growth of the testis, presumably because it inhibits the anterior pituitary gland from producing the gonadotrophic hormones, by the feed-back mechanism.

Testosterone also appears to be responsible for behaviour changes in animals, and administration of the hormone to sexually immature males results in the appearance of male breeding behaviour. In some species even the female will show a masculine pattern of breeding or sexual behaviour when given testosterone. The aggressiveness of many male mammals can be promoted by giving testosterone. In higher primates, including man, sexual behaviour cannot be controlled so simply, and cultural factors play a much more important role in the determination of the direction of sexual impulses.

HORMONES AND REPRODUCTION IN THE FEMALE

In the mature female mammal sexual activity tends to be an intermittent phenomenon during the breeding season, which is the period when mating can occur. The breeding season may consist of several weeks, or months of the year, and there may be more than one breeding season in the year. During the breeding season itself sexual activity is a cyclic phenomenon, with periods of sexual activity or oestrus ('heat') alternating with periods of sexual inactivity. These cycles of activity stop of course as soon as a pregnancy is started. In some animals, including primates and rodents, these oestrous cycles are continuous throughout the sexual life of the animal and are not restricted to breeding seasons; these animals are said to experience poly-oestrus. In man, because of the unusual feature of menstruation of the oestrous cycles are known as menstrual cycles; but the menstrual cycle is fundamentally similar to the oestrous, cycle of other mammals. The variations in breeding activity are best illustrated by referring to particular examples:

Horse

The mating season of the horse is between March and August, although some breeds will mate in the autumn and winter in England. During the mating **season,** if pregnancy does not occur then there is a regular occurrence of oestrus, each period of oestrus lasting about 20 days. The horse is said to be seasonally poly-oestrous.

Dog

The domestic dog has two breeding seasons in the year, in late Winter and early spring and in the Autumn. During each season there is only one period of oestrous, and" the dog is said to be monoestrous.

Golden Hamster

This rodent is polyoestrous,-coming -into heat at all times of the year, the oestrous cycles recurring every ten days, each cycle lasting about four days.

Roe Deer

Like the dog the Roe Deer is monoestrous. The breeding season is in July and August during which there is only one period of oestrous.

The Oestrous Cycle

During the oestrous cycle there are wide-spread changes in the structure and behaviour of the female; all the changes that occur are under the control of hormones, produced mainly by the anterior pituitary gland and the ovary. The aim of these changes is to mature an ovum and prepare the uterus to receive and nurture the ovum if it is fertilized by a sperm.

In the early phase of an oestrous cycle the ovary is activated by the secretion of follicle-stimulating hormone (F.S.H.) from the anterior pituitary gland. The ovary responds by the progressive growth of one or more Graffian follicles, depending upon the species of mammal. Small amounts of luteinizing hormone (L.H. is said to have a synergistic effect. The ripening Graafian follicles secrete a steroid sex hormone called oestradiol which has widespread effects upon the secondary sex organs, and also has an effect on the secretory activity of the pituitary gland. Oestradiol is only one of a number of substances isolated from the ovary, blood and urine of the female mammal, all of which have some effect on the sex organs; the name oestrogen have beenused to describe this class of substances, although oestradiol is the most potent naturally occurring oestrogen.

Oestradiol stimulates the growth of the uterus; the myometriuni increases in, thickness because of growth of its individual cells, its vascularity increases and the endometrium thickens, becoming more vascular as its secretory glands grow in length. The Fallopian tubes and vagina are also stimulated. In the sexually inactive phase or anoestrus, the epithelial lining of the vagina is thin, only one or two cells thick, but after stimulation by oestrogen the epithelium thickens and cornifies, and flattened eornified squames appear in the vaginal secretions. The mammary glands also increase in size under the influence of oestrogen, which stimulates the growth of hė duct system.

The rising level of oestradiol in the blood has important effects upon the pituitary gland. It inhabits the formation of F.S.H. and at the same time stimulates the production of further amounts of L.H. which brings about ovulation and the formation of the corpus luteum. This first phase of the oestrous cycle, as described above, is called the *follicular phase,* because of the growth of the Graafian follicles, or the oestrogen phase, because of the importance of this hormone in this part of the cycle. In the uterus the follicular phase is associated with growth, both of the myometrium and endometrium, and this phase of uterine change is called the proliferative phase. The follicular phase finishes at ovulation when the egg escapes from the ovary and the remains of the Graffian follicles grow to produce special glandular structures under the influence of the pituitary luteinizing hormone (L.H.), called *corpora lutea.*

We now enter the *luteal phase* of the oestrous cycle. This is often called the progesterone phase, because of the importance of this hormone at this time. The corpora lutea are large yellow pigmented bodies studded in the ovarian cortex. They are formed by the proliferation and growth of cells in the wall of the ruptured Graafian follicle. Under the influence of another pituitary hormone called luteotrophin or lactogenic hormone the corpora lutea secrete a hormone called progesterone, in addition to small amounts of oestradioL Pugesterone produces further changes in the endometrium of the uterus the increase in thickness and vascularity of the-uterus progresses as the glands of the endometrium become tortuous and begin to persecretions into the cavity of the uterus. Because of these glandular changes this phase of uterine activity is called the *secretary phase.* Progesterone also has effects on the mammary glands . which haw already been primed by the effects of oestradiol; now, glandular element begin to appear around the ends of the duct systems of the mamma glands.

We have seen that the time of maximal oestradiol activity is at the time of ovulation and after this time the level of oestrogen production gradually falls, as progesterone comes to play a more important par in the cycle. It is at the time of maximal oestradiol production that the female mammal is most willing to receive the male and this is the true period of 'heat' Mating and fertilization usually occur about this time. If fertilization of an ovum does not occur then corpora lutea gradually disintegrate and retrogressive changes occur in the secondary sex organs. The falling level of blood oestrogen in the luteal phase the cycle, together with some inhibitory effect of progesterone on the anterior pituitary gland, are responsible for a gradual decline in the production of L.H. and therefore the corpora lutea degenerate.

Menstrual Cycle

In the human female when the corpus luteum disintegrates at the end of the luteal phase of the cycle, there is a complete breakdown of the hypertrophied endometrium; blood and broken-down tissues are discharged from the vagina and this constitutes menstruation. In the human the menstrual cycle lasts about 28 days. In the first half of the cycle there is the follicular phase, culminating in ovulation at about the foul teenth day. In the second half of the cycle, the luteal phase, the corpus luteum is formed and the endometrium enters the secretory phase, and in the absence of fertilization the luteal phase is ended by the appearance of the menstrual flow. Following menstruation only fragments of endometrium are remaining and these lie in crypts in the myometrium. From these fragments the entire endometrium is reformed during the proliferative phase of the next cycle. The primate endometrium undergoes this almost complete. breakdown because of a peculiarity in its blood supply. When the supply of progesterone is waning as the corpus luteum degenerates at the end of the luteal phase, certain spiral. arteries of the endometrium go into such intense spasm that the tissues supplied by them die and undergo degenerative changes. The whole of the dead endometrium is then sloughed off from the uterine wall together with blood.

The Oestrous Cycle and Pregnancy

If during the luteal phase of the oestrous cycle an ovum is fertilized and settles in the uterine cavity then the retrogressive changes in the secondary sex organs do not occur, nor does the corpus luteum degenerate. We have seen that in the absence of pregnancy the corpus luteum degenerates; it does this because of the decline in the production of pituitary gonadotrophic hormones. As soon as there is a union established between the fertilized ovum and the uterine wall increasing amounts of gonadotrophins appear in the maternal blood and this maintains the structure and function of the corpus luteum. The gonadotrophic hormone is produced by the placenta, the organ which unites the mother and foetus, and through which is receives its nourishment. The placenta also produces large amounts of oestrogen and progesterone and gradually replaces large amounts of oestrogen and progesterone and gradually replaces the corpus luteum as a source of these hormones, so that later in the pregnancy one can remove both ovaries from the female mammal without disturbing the pregnancy: The function of the large amounts of sex hormones produced by the placenta is to promote further growth of the uterus, vagina and mammary glands.

During pregnancy, growth of further Graafian follicles and ovulation is prevented because the large amounts of oestrogen produced by the placenta inhibit the anterior pituitary gland from producing follicle-stimulating hormone.

Reproduction and the Environment

The reasons why animals tend to breed at relatively restricted times of the year have been studied and classified under two headings, internal physiological mechanisms on the one hand, and external environmental factors on the other. An internal physiological 'clock' cannot be the sole factor in determining periodic breeding; seasonal breeding has an adaptive significance in that young are produced at favourable times of the year, and obviously a rigid internal mechanism would, through the course of time, fail to adapt the animal to changing climatic conditions.

It appears that animals have become adapted to respond to certain environmental factors which herald the oncoming favourable season. In temperate latitudes many animals have as it were, harnessed their breeding behaviour to the length of day and respond to an increase in day length by development of the sex organs, so that the young will develop in a favourable season with warmth and an adequate food supply. The first study of the effect of light on sexual cycles was made on the Canadian bunting, a bird which normally breeds in Spring. By giving these birds extra periods of light in the Autumn it was possible to cause development of the testes, even at a time when the temperature animals, reptiles, fish, birds and mammals it has been possible to cause the development of the sex organs out of the breeding season. The light exerts its effect by way of the eyes and connections with the hypothalamus and anterior pituitary gland. Not only is the increase in day length of importance but in some autumn breeding animals the shortening days may also act as the stimulus to sexual development.

When some animals with fixed breeding seasons in temperate latitudes are transferred from the Southern to the Northern Hemisphere, after a However, animals imported from tropical countries tend to continue differences in daylight to which the species inhabiting temperate zones respond. Thus may be due to the fact that owing to the comparative uniformity of conditions in their own countries they have never acquired the capacity to respond to variations in light intensity or duration, characteristic of many animals living under seasonally changing conditions. Thus Java deer when imported to England continue to produce young in late Autumn, as they are believed to do in Java, a condition which is abnormal for any deer inhabiting temperate countries. But even in tropical climates animals may have special breeding seasons; if reproduction were continued throughout the year the competition for food for the young might be too intense for survival of the species. A further advantage of seasonal breeding may be the synchronization of the male and female sexual cycles by the fact that they are both adapted to the same environmental factor. What these environmental factors are in tropical climates in uncertain. That breeding cycles in some tropical animals are harnessed to some environmental factor and not dependent upon an internal rhythm seems certain; a tropical insectivorous bat lives throughout the period of daylight until about ten minutes before sunset in dark and almost thermo-static caves and yet it was found that in one year of observation no pregnancies occurred until a few days at the beginning of September. It seems impossible that such synchronization of the sexual cycles of such a group of bats could be achieved by an internal physiological mechanism.

The homiothermic vertebrates (warm blooded) have become almost independent of the temperature of their surroundings and are able to carry out breeding at any time of the year. They do, however, tend to breed in Spring in temperate latitudes to ensure a favourable environment with adequate food for the young. But even in homiothermic animals temperature may play some role in the timing of breeding behaviour. If Autumn weather is warm and food supply is good, sexual behaviour in the robin and other birds may be pronounced and may even lead to reproduction in a few birds, although they normally only breed in Spring. In this Autumn breeding phase sexual activity is developing hen daylight is decreasing.

We have now seen something of the way in which sexual' cycles are controlled by the endocrine organs, and the way in which the endocrine organs, by way of the hypothalamus of the brain, are synchronized with external environmental factors so that the young are cared for under favourable conditions.

THE PLACENTA

The uterus has been preparing to receive a fertilized ovum throughout the oestrous cycle. In the follicular phase of the cycle there is growth both of the myometrium and endometrium and an increase in the blood supply of the uterus. The endometrium is thickened, new blood vessels grow and its glands increase in length. In the luteal phase of the cycle these changes continue and the endometrium takes up secretary character. The glands become tortuous and their epithelium becomes active, secretions passing into the uterine cavity.

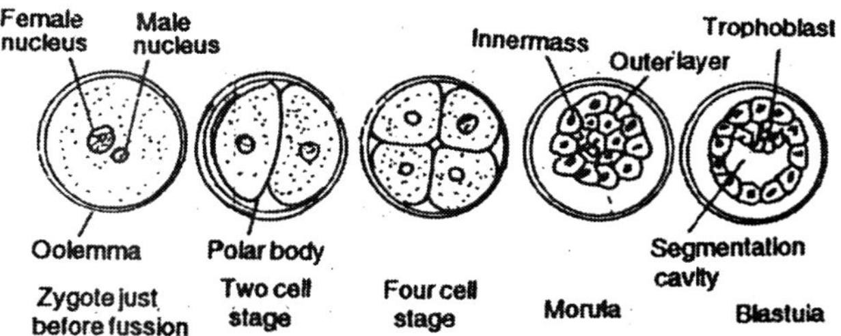

Figure 2.8 : Diagrams showing the early development of the fertilized egg of the mammal up to the blastula stage.

The early development of the fertilized egg occurs during its passage through the Fallopian tube, since fertilization is usually achieved high up in the Fallopian tube. The fertilized egg or zygote undergoes divisions to produce a ball of cells called the morula. Then morula than differentiates into an outer layer and an inner cell mass producing the blastula. The cells of the outer layer form the trophoblast or trophoblastic ectoderm. This enters into the formation of the chorion, the outermost, covering of the developing zygote. This outer layer is called *trophoblast* because it enters into the formation of placenta, the organ concerned in nourishing the foetus. The inner cell mass of the blastula contains the cells from which the embryo will develop.

During its passage down the Fallopian tube the nutrition of the zygote is dependent on the small amounts of food stored in the protoplasm. When the morula reaches the uterine cavity it is bathed in the secretions of the uterine glands, and these secretions probably have some nutritive function. In some species the blastocyst lies in the cavity of the uterus, and in contact, by means of its trophoblast, with the endometrium all over its surface. In some other species the blastocyst becomes attached to the endometrium on one surface, and projects freely into the uterine cavity or its other serface. In other types the blastocyst sinks into the endometrium and becomes surrounded completely by maternal tissues; this type of implantation of the blastocyst is found in man and is called the interstitial *type*. In order to understand the varied types of placenta found in mammals it is necessary to examine in some detail the formation of the various foetal membrane which take part in the formation of the placenta. There are four foetal membranes concerned in the adaptation of the foetus to life in the uterus, the amnion and chorion, formed from the original embryonic body wall, and the yolk sac and allantois, parts of the original gut of the embryo.

We had left the development of the embryo at the blastula stage, consisting of an *inner* cell mass and an outer layer, the trophoblast; a two layered vesicle is produced by a growth of endoderm cells around the inner layer of the trophoblast (enclosing the yolk of the egg, when this is present) Only a part of the wallof the blastula is destined to form the embryo and this is called the embryonic area. This embryonic area gradually sinks into the centre of the blastula as it becomes covered over by the amniotic folds. In man the embryo is not covered by the amniotic folds and there is merely a hollowing out of the cells of the embryonic area to produce a cavity, the amnion. The amniotic folds meet and fuse above the embryo and because of the double walled nature of the folds the embryo becomes to be surrounded by two membranes, an outer chorion

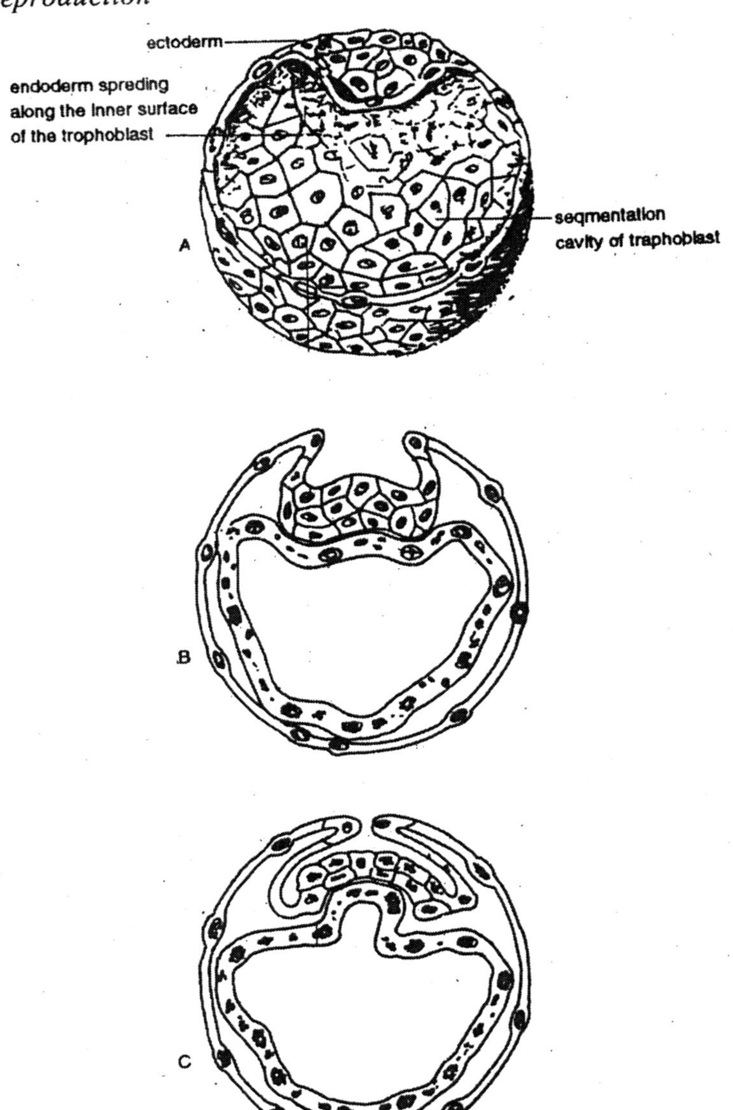

Figure 2.9 : Development of the late blastula. Note the embryonic area gradually sinking into the blastula in A, B and C, gradually becoming covered by the amniotic folds. The endoderm is shown progressively spreading over the inner surface of the trophoblast.

and an inner amnion. The chorion thus becomes the membrane which is in contact with the endometrium of the uterus, and projections called chorionic villi grow out from its surface to make a more intimate contact with the maternal tissues. The amniotic membrane surrounds a fluid filled cavity, the amniotic cavity, in which the embryo floats.

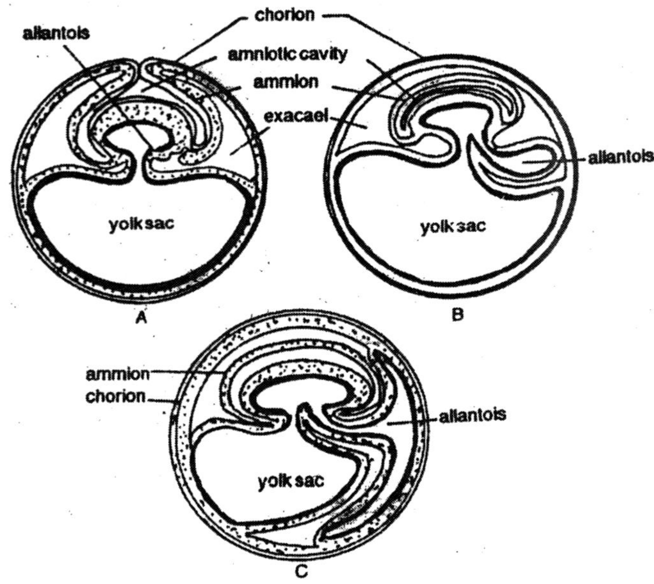

Figure 2.10. Showing the development of extra embryonic structure.

We have seen above how the endoderm grows round the inner surface of the trophoblast to produce a two layered vesicle. In the heavily yolked eggs of birds and reptiles this layer encloses the yolk of the egg and is called the yolk sac. The primitive gut of the embryo, a groove on the under surface of the embryonic area, is open into this yolk sac but later as the gut is folded off it is separated from the yolk sac except for a narrow passage, the yolk sac stalk. Later mesoderm grows out from the embryo between ectoderm and endoderm; this process provides a double layer of mesoderm in the amniotic folds and a layer of mesoderm between the ectoderm and endoderm of the yolk sac. It is in this layer of mesoderm that the embryonic blood vessels develop. In the heavily yolked eggs of birds and reptiles the inner layer of the yolk sac becomes vascular in order to absorb the nourishment from within the yolk sac. But in mammals there is very little nourishment within the yolk sac and it is from outside that it has to look for nourishment. Thus the outer layer of the trophoblast becomes highly developed and supplied with blood vessels from the mesoderm. In primate embryos, including man, the yolk sac is a rudimentary structure only. In Marsupial mammals which have a less developed placenta the yolk sac is a more important organ, and is the major organ for nourishment of the embryo. In rodents the yolk sac becomes partly vascularized; the non vascularized area becomes eroded away so that the cavity of the yolk sac opens into the uterine cavity. The inner wall of the yolk sac is applied directly to the endometrial lining of the uterus, producing what is known as an inverted yolk sac placenta.

In most mammals the chief absorbing surface occurs on a structure called the allantois. This is a blind ended outgrowth of the hind gut carrying mesoderm on its surface as it grows into the extra embryonic coelom. In the eggs of reptiles and birds the allantois functions as a bladder for the storage of excretory products, but it also serves to transport oxygen which diffuses through the shell, through the blood vessels of the allantois to the developing embryo. In mammals the allantois has become progressively more important in the nourishment of the foetus, and blood vessels of the allantois pass into the villi of the chorion which make connection with the lining of the uterus.

The processes which grow out from the trophoblast (i.e. extra-embryonic ectoderm of the chorion) are called trophoblastic or chorionic villi. The villi are connected to the embryonic blood vessels through the yolk sack or allantois, depending upon the species of mammal. Not all of the surface of the trophoblast is covered by villi and the arrangement of the villi varies from species to species. In the horse and the pig the villi are diffusely arranged over the surface of the trophoblast producing a diffuse placenta. In the sheep and cow the villi are localized in patches called cotyledons. producing the cotyledonary placenta. In carnivores the villi are restricted to a band encircling the embryo producing a zonary placenta. In man (also rodents and insectivores) the villi are at first scattered over the whole trophoblast and later become limited to a disc shaped area producing a discoidal placenta.

Types of Placental Union with the Uterine Wall

There is a great variation in the intimacy of contact between the foetal and maternal tissues in the placenta of mammalian species. On the foetal side of the placenta there are three layers of tissue, the chorion, the mesenchymal tissues and the endothelium of the 'capillary vessels. On the maternal side there are uterine secretions, the endometrial epithelium, connective tissues and the endothelium of the blood vessels. There are thus at least six possible layers of tissue separating the foetal and maternal blood. In the diffuse placenta found in the pig and horse these six layers which separate the maternal and foetal blood streams persists, producing what is called an epithelio-chorial placenta.

This description of the placental structure of the pig and the horse give a rather exaggerated picture of the separation of maternal and foetal blood. The foetal cells of the chorion are low cuboidal in shape and foetal capillaries ramify close to the chorion and may actually penetrate between the chorionic cells to form intra-epithelial plexuses. The capillaries at places displace the cuboidal cells and compress their cytoplasm into thinned out plates. Thus over a large part of the placental surface only a very thin plate of epithelial cytoplasm separates, the foetal and maternal blood. In other mammals the barrier between maternal and foetal blood is reduced by the erosion of the layers of the uterine wall by the trophoblastic villi. In the syndesmochorial placenta of cattle and sheep the endometrial epithelium is eroded and the epithelium of the trph oblastic villi is in contact with the uterine connective tissue and in the endotheliochorial placenta of the cat and dog the uterine tissues are further eroded and the epithelium of the trophoblastic villi is separated from the maternal blood only by the endothelium of the maternal capillaries. In insectivores, rodents and man the maternal capillaries are eroded so that the trophoblastic villi are bathed in lakes of maternal blood. The epithelial cells of the chorion are provided with many minute projections called microvilli which

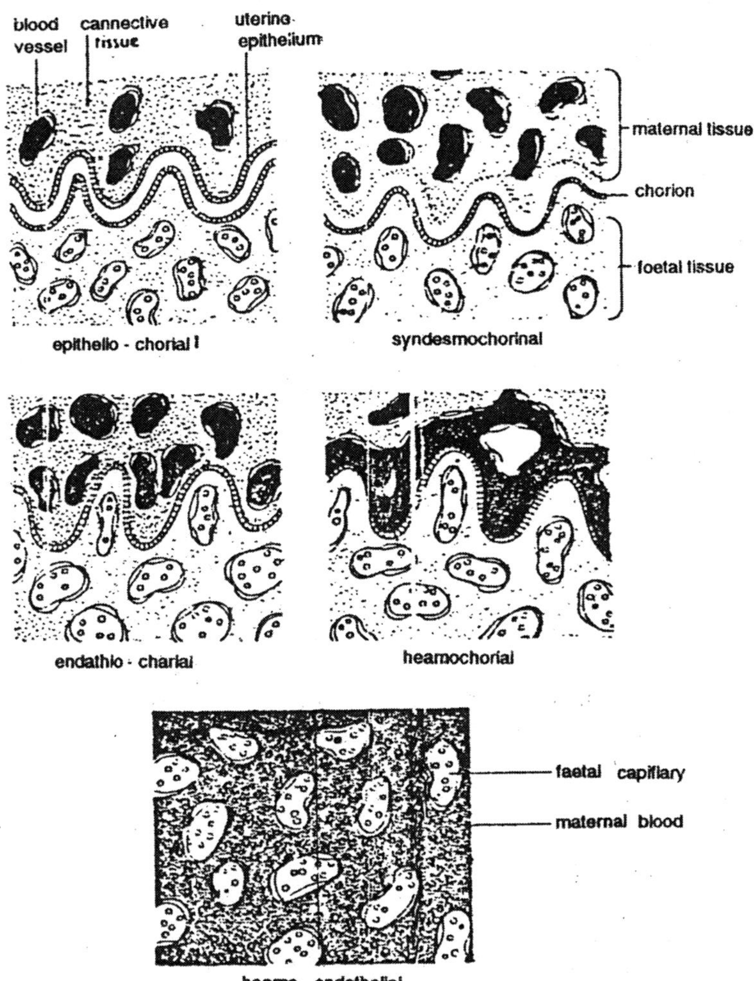

Figure 2.11 : Diagrammatic representation of different placental types.

interdigitate with maternal tissue and provide a very intimate form of contact between the two. The long processes of foetal tissue, called trophoblastic villi, run vertically from the surface of the placenta into maternal tissues. Between these trophoblastic plates are the maternal capillaries with a thick endothelium. These capillary endothelial cells show, in electron micrographs, an interesting structure which may well be related to the way in which some substances are transferred by the placenta between maternal and foetal blood. The cytoplasm of the maternal capillary endothelium is spongy, containing many vacuoles. In the cytoplasm between these vacuoles are mitochondria and Golgi material. The vacuoles are sometimes seen to connect directly with the lumen of the capillary and it seems probable that the vacuoles are formed by invagination of the surface membrane of the cell. Thus substances included in the vacuoles may be transferred by these means from maternal blood to foetal tissues without having to pass through the 'cytoplasm' of the endothelial cells. The basement membrane of these endothelial cells varies a great deal in thickness, being absent in some places so that the plasma membrane of the maternal endothelial cells may be in direct contact with the cells of the foetal trophoblastic villi, The irregular basement membrane may represent erosion by the trophoblastic villi.

In the human placenta the trophoblast consists of an inner layer of cells (Langhans cells) lying next to the foetal connective tissue and blood vessels and an outer sycitial layer which is bathed by the maternal' blood. The Langhans cells become flattened and many disappear as pregnancy advances. The cytoplasm of the syncytium is foamy and many ovoid nuclei are scattered through it. The foaminess of -the cytoplasm is due to the presence of many vacuoles, each with a granular outline, the granules probably being of R.N.A. Synthesis of foetal protein may occur at these sites until the foetal liver takes over this role.

In addition to these smaller vesicles there larger vacuoles, usually near to the outer surface of the syncytium under the microvilli which occur here. These vacuoles may rise by invaginations of the surface membrane of the cell and may be concerned with the 'absorption' of maternal plasma into the cell by the process of pinocytosis. The large number of microvilli increases the surface area of the cell for absorption or secretion.

In the rabbit, guinea pig and rat there is even more intimate contact since the epithelium and mesenchymal tissues of the chorionic villi disappear leaving only the endothelium of the trophoblastic capillaries separating maternal and foetal blood, producing the haemoendothelial placenta.

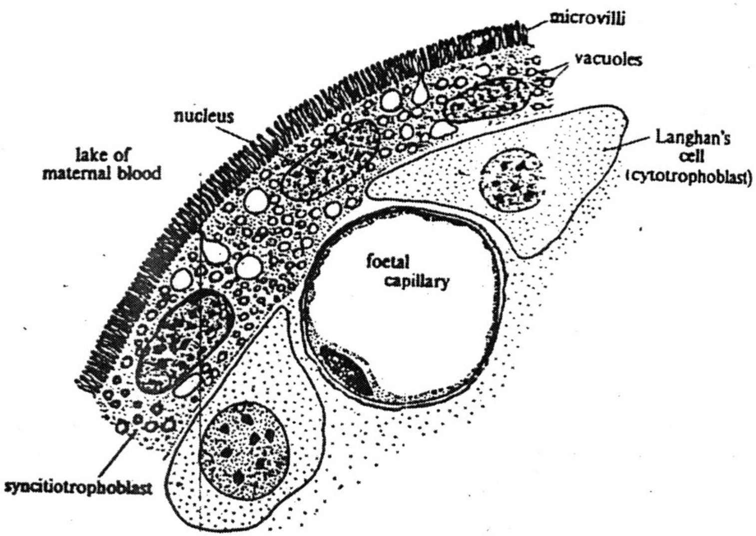

Figure 2.12 : Diagram of a small area of human placenta made from electron micrographs. Note the microvilli of the syncytiotrophoblast projecting into a pool of maternal blood. Some of the vacuoles of the syncitio-trophoblast are shown opening between the bases of the microvilli.

Type of placenta, efficiency and length of gestation period. In the ungulates with epithelio-chorial or syndesmochorial placentae the barrier between maternal and foetal circulations is much greater than in the other placental types. Nutrition of the foetus in ungulates is assisted by secretions of the uterine glands called uterine milk It is difficult to correlate the efficiency of the placenta with the type of placenta; even in one species the placental structure is not constant throughout development, and in some species e.g. rat, there are additional structures to the chorio-allantoic placenta in the form of the yolk sac placenta.

However there is a general tendency for animals possessing a more highly developed placenta (i.e. fewer layers) to have a shorter gestation period, although the young are born in a helpless condition. This feature is illustrated in the following list of gestation periods :

Mouse gestation period 19 days. Haemoendothelial placenta. Young very immature at birth.

Rat gestation period 21 days. Haemoendothelial placenta.

Dog gestation period 63 days. Endothelio-chorial placenta. Young helpless and blind.

Cat gestation period 65 days. Endothelio-chorial placenta. Young are helpless and blind.

In the above examples the young are in a very similar condition at birth, although the gestation period of the cat and dog is three times as long as that of the mouse and rat. This difference is explained on the grounds of a more efficient placenta is rats and mice.

SUMMARY OF TYPES OF PLACENTA

Placental type	*Epithelia chorial*	*Syndesmo-chorial*	*Endothelio-chorial*	*Haemo-chorial*	*Haemo Endothelial*
Maternal tissues					
Endothelium	+	+	+	–	–
Connective tissue	+	+	–	–	–
Epithelium	+	–	–	–	–
Foetal tissues					
Chorionic epithelium	+	+	+	+	–
Mesenchyme	+	+	+	+	–
Endothelium	+	+	+	+	+
Examples	Horse Pig	Cattle Sheep	Dog, Cat	Insectivores Man, Lower Rodents	Rat Rabbit

+ indicates presence
– indicates absence

Guinea Pig. gestation period approx. 68 days. The young are born active with eyes open. There is an accessory yolk sac placenta throughout the gestation period. The additional period of gestation compared to the mouse and rat is used to further the development of the young.

Pig. gestation period 119 days. Epitheliochorial placenta. This is a primitive type of placenta and a long gestation period is necessary.

Thus it is seen that the more primitive type of placenta is associated with a longer gestation period, the guinea pig being an exception to this general rule perhaps because the young are so well developed by the time they are born.

PHYSIOLOGY OF THE FOETUS

The Placenta and Metabolism of the Foetus

Through the placenta the foetus obtains its nourishment; water, carbohydrates, proteins, fats, mineral salts, vitamins and oxygen, and through the placenta pass the waste products of the foetus, including carbon-dioxide and urea, to be excreted eventually by the lungs and kidneys of the mother. The blood vessels of the foetus

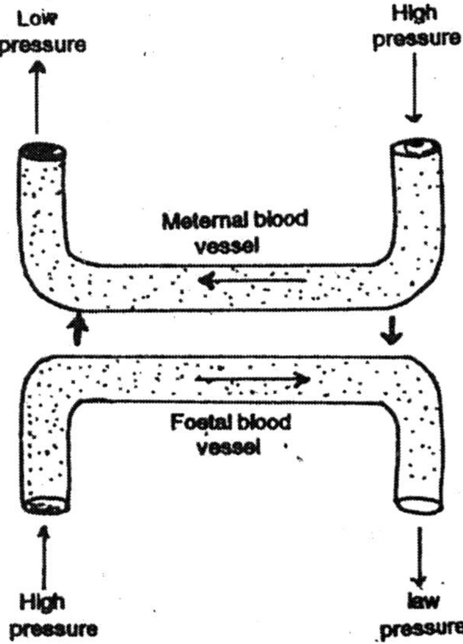

Figure 2.13 : Diagram to illustrate the principle of the arrangement of blood vessels in the placenta. The two short thick arrows indicate the direction of the exchange of materials.

and mother do not join one another and all these substances must diffuse across the barrier between the two sets of blood vessels. There are three devices in the placenta which promote the passage of materials across the barrier between the two sets of vessels. We have already seen two of these devices: first the large surface area of the placenta provided by the branching villi and secondly the reduction in the thickness of the placental barrier with varies from species to species. The third device consists in the arrangement of maternal and foetal blood vessels so that the maternal blood vessels containing blood at relatively high pressure are first adjacent to foetal vessels containing blood at low pressures, whilst foetal blood vessels containing blood at relatively high pressure lie adjacent to maternal vessels containing blood at low pressure: this arrangement encourages an efficient transfer of substances between the two series of blood vessel.

Respiration of the Foetus

The need for an increasing supply of oxygen by the foetus is somewhat anticipated by the maternal part of the placenta in that the maternal vascular bed in the uterus grows more rapidly than the foetal contribution to the blood vessels of the placenta. Thus, at an early stage in gestation the maternal blood leaving the placenta still contains large amounts of oxygen. As the gestation period progresses the maternal blood leaving the placenta becomes progressively more de-oxygenated; the placenta reaches its maximum size at a time when the foetus is still growing rapidly, and at the end of the gestation period the supply of oxygen is critical.

The foetus is adapted in two main ways to obtain oxygen efficiently from the placenta. First there is the special arrangement of foetal blood vessel which has already been described above. Secondly, in many species foetal haemoglobin has somewhat different properties from adult haemoglobin, in that it takes up oxygen with a greater avidity and gives it up less readily. foetal haemoglobin can take up oxygen at partial pressures at which maternal haemoglobin would give up oxygen. The foetus is a relatively inactive creature and can exist with haemoglobin which retains its oxygen more avidly than does adult haemoglobin. But once the animal is born it may need to use oxygen as rapidly as does an active adult, particularly in those species such as the horse where the young are very active at birth. It has been found that in some animals, e.g. the goat, there is a gradual alteration of the properties of foetal haemoglobin produced towards the end of the gestation period to approximately those of the adult, so that the animal will be better adapted to an active life breathing air.

Carbohydrate Metabolism

As in the adult, the foetus uses glucose as an important source of energy. Glucose can pass across the placenta. from mother to foetus in all animals. There may be persistent differences between the blood sugar level of the foetus and mother which may imply that the transfer of glucose across the placental membranes is not merely one of diffusion. In most mammals the foetal blood sugar tends to be lower than that of the mother but in some epitheliochorial placentae, e.g. pig, the blood sugar is higher on the foetal side.

In early foetal life the maternal side of the placenta acts as a temporary liver for the foetus in that it holds stores of glycogen at a time when the foetal live contains little or no glycogen. Later in the gestation period when the foetal liver has developed stores of glycogen, less in stored in the placenta. From experiments it has been found that the foetus exercises a strict glycogen economy, drawing upon glycogen stores only in emergencies.

Lipid Metabolism

The foetus usually has good stores of fat. This may come from several sources. It may be able to synthesize fat from carbohydrates and amino acids. Some lipids pass across the placenta and become available for the synthesis of fat, but the mechanism of transfer across the placental membranes is not understood. Most of the fat fed to the mother and stored in her tissues does not pass unchanged across the placental barrier. If the food fat is stained with the dye Sudan III than the mother's stores of fat are intensely stained red, although none appears in the foetus. Further evidence which indicates that fat is not transferred directly across the placental membranes is the different chemical constitution of foetal and maternal fat.

Protein Metabolism

Foetal protein may be derived directly by transfer of intact protein molecules from the mother or by transfer of amino acids. Antibody proteins (gamma globulins) are known to be transferred intact from mother to foetus in some species (e.g. rabbit) and this confers on the foetus a passive type of immunity. In other species much or all of the antibody proteins are received by way of the colostrum, the first pale secretion of the mammary glands of the mother. Amino acids probably do not pass across the placenta by simple diffusion and active transport mechanism may be involved.

Endocrine Function of the Placenta

In addition to serving the nutrition of the foetus the placenta is an endocrine organ producing several hormones in large quantities. Gonadotrophin is produced from the chorion and is called chorionic gonadotrophin. In the human this hormone is produced in such large quantities by the end of the second month of pregnancy that a specimen of urine, in which the hormone is excreted, will cause the growth of the ovaries and ovulation when injected into a sexually immature mouse. Modifications of this effect are used in the early diagnosis of pregnancy. When the urine is injected into a female toad called Xenopus, eggs are shed into the surrounding water within 24 hours. Even mature male amphibians have been used in pregnancy tests and injections of chorionic gonadotrophin are followed by a release of sperms, which have to be identified in the urine microscopically. In addition to gonadotrophin the placenta also produces large amounts of oestrogen and progesterone. The exact significance of these various hormones is not well understood but they undoubtedly play their part in the development and maintenance of the sex organs during the pregnancy and prepare the mammary glands for their function after the birth of the young. The high local concentration of progesterone at the placental site, by increasing the membrane potential of the uterine smooth muscle cells renders them less excitable. Thus the uterus is mechanically inactive and the foetus is safeguarded.

The circulation of the Foetus and Modifications at Birth

The circulation is adapted to life in utero, in which the placenta is the organ of nutrition and respiration; the foetal lungs and gastro-intestinal tract do not function as the respiratory and nutritive organs in utero and their blood supply is correspondingly small. But immediately at birth the placenta ceases to have these functions and there must be a rapid adjustment of the organism to an independent life. The lungs are rapidly converted from semi-solid organs with a small blood supply into air filled organs with a.large blood supply and are responsible for the gaseous exchange of the whole organism. The gastro-intestinal tract takes over the nutritive functions of the placenta as the young animal begins to feed. In order to make these transformations possible there must be special devices within the cardio-vascular system of the foetus so that blood which once went to the placenta can now be diverted to the lungs and gastro-intestinal tract.

Foetal oxygenated blood is returned from the placenta by way of the umbilical vein which passes to the liver where it joins the hepatic portal vein. Some of the oxygenated blood goes to the liver but most of it is shunted away through a connection of the umbilical vein with the inferior vena cava called the ductus venosus, a feature only of the foetal circulation. In the inferior vena cava the oxygenated blood mixes with venous blood draining from the lower part of the body. When this blood reaches the heart most of it is shunted from the right auricle through an opening in the septum between the two auricles, into the left auricle. The opening in the septum, called the foramen ovale, is a special feature of the foetal heart and it censures that the oxygenated blood returning from the placenta is diverted away from the right ventricle and lungs to supply the head (i.e. central nervous system) and upper limbs of the foetus by way of the left ventricle and aorta. This flow of blood has been confirmed using X-ray studies of the foetus after radio-opaque material has been injected into the circulation. The blood is diverted from the right auricle into the left auricle by a valvular arrangement in the wall of the right auricle.

The venous blood which drains into the right auricle from the head and neck of the foetus by way of the superior vena cava passes through the right auricle into the right ventricle, from which it passes out into the pulmonary artery. But since the foetal lungs are not functioning as organs of gaseous exchange, only a small amount of blood is needed, sufficient to meet the metabolic needs of the tissues in the lung and much of the blood in the pulmonary artery passes directly into the aorta by way of a connection called the ductus arteriosus. This blood supplies the lower part of the body, and much of the it passes into the umbilical arteries (branches of the internal iliac arteries) supplying the placenta with deoxygenated blood.

There are thus four special features of the foetal circulation:

1. The placental circulation.
2. The ductus venosus which shunts blood from the umbilical vein away from the liver into the inferior vena cava.
3. The foramen ovale through which oxygenated blood returning to to the heart via the inferior vena cava passes into the left auricle and so to supply the upper part of the body.
4. The ductus arteriosus which shunts blood from the pulmonary arch into the aorta, so by-passing the lungs.

It will be seen that the right ventricle of the foetus, pumping blood to the lungs and to the lower part of th body and placenta, performs more work than the left ventricle. This position is reversed after birth.

At birth there are dramatic changes in this circulation. The umbilical vessels become increasingly irritable towards the end of the gestation period and with the physical stimuli of birth these vessels go into spasm, thus excluding the placental circulation from the foetus. This means that when the mother bits through the umbilical cord to free the young animal it does not bleed to death. The ductus venosus also goes into spasm so that blood in the hepatic portal vein now has to pass through the tissues of the liver and cannot be directly diverted into the inferior vena cava. With the expansion of the lungs at birth larger amounts of blood pass into the lungs from the plumonar arch, and the wall of the ductus arteriosus contracts so that blood in the pulmonary arch can no longer be diverted into the aorta. With the increased supply of blood to the lungs there is an increasing amount of blood returning to the left auricle and the pressure of blood in the left auricle rises. This rise in pressure pushes a loose flap of tissue against the foramen ovale, closing the connection between the left and right auricles. The adult type of circulation is now achieved. The ductus arteriosus, ductus venosus and umbilical vessels, initially closed by muscular spasm are gradually permanently obliterated as fibrous tissue grows in the lumen of the vessel, and the flap of tissue which is closing the foramen ovale gradually fuses with the septum.

Parturition

At the end of the gestation period parturition occurs when the young are expelled from the uterus into the outside world by means of powerful intermittent contractions of the myometrium. What initiates the process of parturition is not known but it has been suggested that the declining production of the hormone progesterone, which has an inhibitory effect on uterine motility, sensitizes the uterus to the posterior pituitary hormone oxytocin, which stimulates intermittent powerful contractions of the uterus. In cases of slow labour associated with weak uterine contractions, injections of oxytocin are used to produce more powerful uterine contractions.

In those species in which there is an intimate mingling of foetal and maternal tissues in the placenta the uterine contractions not only propel the young through the birth canal but they serve to separate placenta from the uterine wall, and after birth of the young the persistent contraction of the uterine muscles serve to close the blood vessels which have been torn upon during the separation of the placenta.

Lactation

The mammary glands have been prepared for their function during the pregnancy by the effect of the sex hormone produced by the placenta. As mentioned previously oestrogen causes a growth of the duct system of the glands and progesterone initiates the development of the glandular elements around the ducts. Btu the glands are unable to produce milk during pregnancy because the large amounts of oestrogen inhibit the pituitary gland from producing a hormone vital for lactation, the lactogenic hormone.

Towards the end of pregnancy the oestrogen production by the placenta gradually falls, and after parturition, when the placenta is expelled, the level of oestrogen in the body falls still further Lactogenic hormone is now produced by the anterior pituitary gland and lactation starts soon after parturition. The first pale secretions called colostrum are rich in proteins, and in some species they contain antibodies, which serve to protect the young for some months until they have been broken down by the body.

During the lactation period milk is secreted by the glandular cells of the mammary gland continuously. Milk, however, only passes along the larger ducts that open onto the nipple during suckling. When the infant mammal grasps the nipple in its mouth stimulation of sense organs in the nipple cause a reflex liberation of the hormone oxytocin from the posterior pituitary gland. This hormone reaches the mammary gland by way of the blood stream. On reaching the mammary gland the hormone stimulates certain contractile cells surrounding the glandular cells causing an emptying of milk into the larger ducts within open onto the nipple. Milk now appears from pores on the nipple and the young, by compressing the dilated ducts which underly the nipple increase the flow of milk.

3

The Nucleus in Development

A living cell is a partnership between the nucleus and the cytoplasm. Now, although this concept has its limitations, it will serve as a basis on which to divide our subject matter. This chapter concerns the nucleus. The remainder of the book deals mainly with the cytoplasm.

The nucleus, with its store of hereditary factors, the genes, is the conservative member of the partnership. It controls the syntheses which go on in the cell. Hence it directs that most amazing of all syntheses, the course of development. Hence also it determines the species and individual , characteristics which are the outcome of development. On the other hand, it is primarily the cytoplasm which develops and is the progressive member of the partnership. It begins in relative simplicity and then, by differentiation and molding, becomes the mature organism in all of its structural and functional complexity.

Both the nucleus and the cytoplasm are needed for the continued life and functioning of the cell. Many experiments have been performed in which the nucleus of a cell has been removed. For example, when an amoeba is cut into two parts, the part containing the nucleus heals and continues to live. The part which lacks a nucleus may carry on for a time, but ultimately it wears itself out and perishes, for it cannot continue to rebuild its substance. It is reported that a sea urchin egg from which the nucleus has been removed may cleave several times, but it will not continue to develop.

It is even more obvious that a nucleus cannot live and carry on its functions without the cytoplasm. In particular, the nucleus depends upon the cytoplasm for its raw material and energy. It is probable that only in the cytoplasm do the oxidations take place by which energyrich molecules (ATP) are synthesized.

THE LIFE CYCLE OF THE NUCLEUS

The story of the nucleus as it is revealed by the microscope is a repetitive story, yet one that is full of interest. It begins at fertilization when the chromosomes of the two germ cells (gametes), an egg and a sperm, come together in the nucleus of the fertilized egg or zygote. The story ends and is ready to start again, at least as far as the contribution of the individual to the race is concerned, when the new individual has matured and has produced gametes of its own. Between fertilization and the maturing of the germ cells, the story of the nucleus is one of alternating growth and division by mitosis.

Fertilization

From the standpoint of the nucleus, fertilization is the coming together of the chromosomes of the egg and the chromosomes of the sperm in the same nucleus, the zygotic nucleus. One set of chromosomes, the haploid number, is present in the egg. Another set is present in the sperm. This makes two sets, the diploid number, present in the zygote. In man the haploid number is 23. Hence the human diploid number is 46.

Each chromosome of a haploid set has its own individuality; that is, it is of its own kind. It becomes visible at the beginning of cell division and disappears at the close of cell division in its own characteristic manner. Very significantly, its genes are arranged within it in a definite sequence. Moreover, it affects development in its own specific way. Each chromosome of a diploid set usually has a mate. Yet, although the two mates are within the same nucleus, with rare exceptions each remains independent of the other throughout the cell divisions (mitoses) of development.

Mitosis

The coming together of the male and female nuclei in fertilization is followed *by mitosis,* the process by which one cell becomes two cells in so complicated but precise a manner that each daughter cell possesses just the same two sets of chromosomes that the mother cell possessed. Several phases of the process are now described.

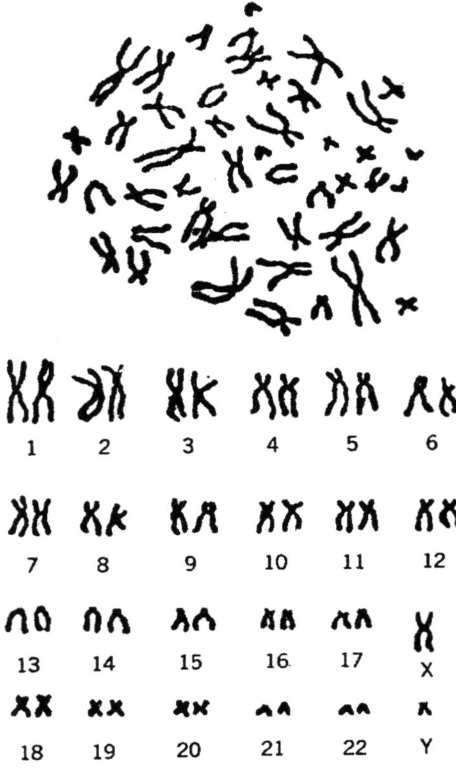

Figure 3.1 : The 46 chromosomes of man, seen in polar view at the metaphase of a somatic mitosis.

Interphase

Let us begin our account with the interphase, that is, with a cell that is not dividing. Such a cell has commonly been called a "resting cell." Actually it is a busy cell doing everything which a cell does except divide. It is during the interphase that the genes self-duplicate and carry on their function of supervising syntheses.

Following self-duplication, the chromosomes continue to be stretched out. Usually they are so attenuated that their boundaries cannot be seen with the light microscope. The extended chromosomes are able to carry on their chemical functions in an efficient manner.

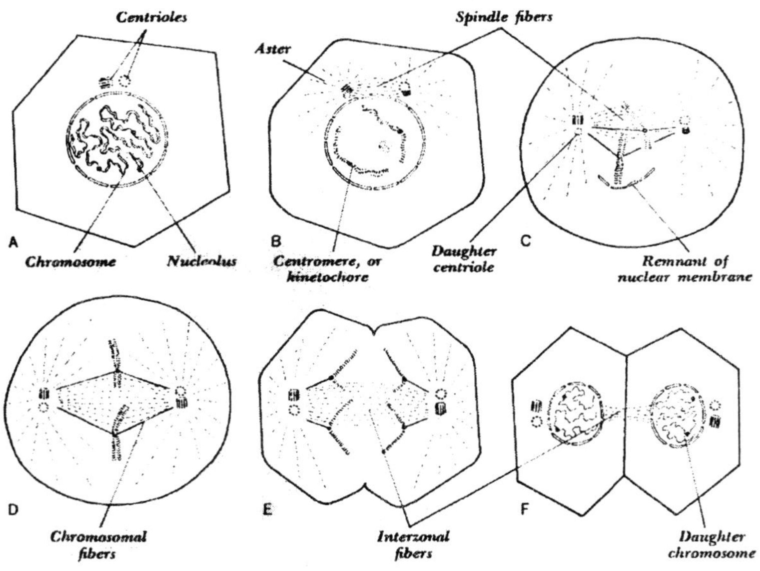

Figure 3.2 : Diagram of mitosis showing one pair of chromosomes. A. Interphase after replication. B. Early prophase. C. Late prophase. D. Metaphase. E. Anaphase. (The telophase is not shown.) F. Early interphase before the chromatids, or "daughter chromosomes" have replicated. The centrioles have been drawn larger in order to show their structure. Replication of the centrioles is shown here beginning in the early prophase, but in some cases it begins as early as the preceding anaphase.

"Nuclear sap" or karyoplasm fills the interstices between the chromosomes. One or more rounded bodies, the *nucleoli,* are usually present in the interphase nucleus. The visible boundary between the nucleus and the surrounding cytoplasm is the *nuclear membrane.* In the cytoplasm adjacent to the nucleus there is a body, the central body, which consists of two granules or, after each granule has replicated, of two pairs of granules, the *centrioles.*

Prophase

The events in the nucleus and in the cytoplasm which lead to the division of the cell constitute the prophase. The following is a generalized account:

1. In the cytoplasm the pairs of centrioles move apart, and a sphere of gel rays, the *aster,* grows around each pair. Together the two asters constitute the *amphiaster.* It is of interest that the centrioles are short, cylindrical bodies, each consisting of nine rods (or double or triple rods). In this they are structurally comparable to a cilium or flagellum. (They lack the two central tubules of a cilium.) In a typical case, the centrioles of a pair lie at right angles to each other and to the axis through the pairs.

2. The chromosomes condense into tight spiral coils. Under the light microscope these appear to be rods. But when the material is suitably fixed and stained, each chromosome can be seen to be a double structure, each half of which is a chromatid.

3. A fusiform body, the *mitotic spindle,* now forms between the pairs of centrioles, sometimes in the cytoplasm, sometimes in the substance of the nucleus. The spindle's poles are anchored in the region of the centrioles of the cytoplasm. The electron microscope shows that it consists of hollow fibrils (microtubules). It elongates as the centrioles move apart. Possibly this is a result of imbibition of water, or incorporation of other molecules, or change in the shape of its constituent proteins. The amphiaster and the spindle together form the "achromatic figure," so-called because it does not stain deeply with basic dyes.

4. According to some accounts, chromosomal fibers grow out from particular locations in each chromosome until they reach the poles of the spindle. The locations from which they grow out are known as *centromeres* or kinetochores. The chromosomal fibers seem to consist of a contractile fibrous gel.

5. Toward the end of the prophase, the nuclear membrane disappears.

6. The nucleolus also disintegrates, and its substance, along with the nuclear sap, passes into the cytoplasm.

Metaphase

The chromosomal fibers appear to contract and pull the chromosomes to the equator of the spindle (prometaphase). When thus spread out and seen in polar view, their number can be counted.

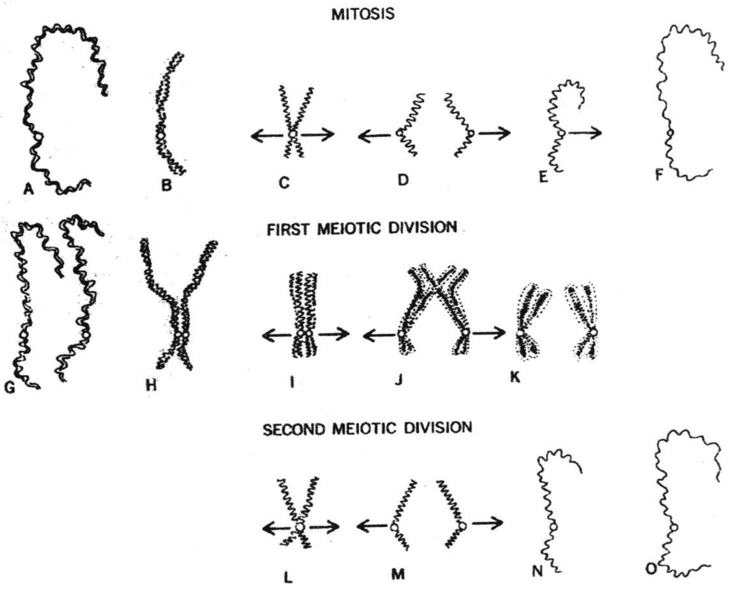

Figure 3.3 : The behavior of chromosomes in mitosis and meiosis. A to* F. *Mitosis, illustrated by a single chromosome: A, interphase after replication has taken place; B, prophase; C, metaphase; D, anaphase; E, telophase;* F, *interphase before replication has occurred. G to* K. *First meiotic division, illustrated by a homologous pair of chromosomes: G, leptotene stage; H, synaptene (zygotene) stage; 1, pachytene stage; J, diplotene (at which time "crossing over" between adjacent segments of homologous chromosomes may take place) followed by diakinesis;* K, *anaphase.* L *to* O. *Second meiotic division, illustrated by one chromosome:* L, *metaphase; M, anaphase; N, telophase;* O, *interphase.

Anaphase

Continuing, or resuming, what appears to be contraction, the chromosomal fibers seem to drag each chromatid toward one pole of the spindle. Actually two processes may be at work: a continuing enlargement and elongation of the spindle and a shortening of the chromosomal fibers. Since the fibers shorten without becoming thicker, the process probably involves the removal of water or other molecules from the fibers. As the chromatids move apart, they are often Vor J-shaped depending on the location of the centromeres to which the chromosomal fibers are attached. A band of "interzonal fibers" is often seen for a time after the separation has been accomplished, connecting the chromosomes which have pulled apart, and often including a remnant of the spindle.

Telophase

Having reached the poles of the spindle, the chromatids swell and disappear from view, as seen with the light microscope. Although usually called "daughter chromosomes," they are actually single chromatids. The nuclear membrane reappears, and shortly one or more nucleoli re-form. Of the mitotic apparatus, little more than the centrioles of the cytoplasm remain.

Interphase

The daughter cells have now returned to the interphase, or "resting stage," except for one feature: each chromosome still consists of only one chromatid. Investigations by a staining technique known as the Feulgen reaction (specific for DNA) have made it clear that the restoration of chromosomal material takes place during the interphase. At that time each chromatid becomes two chromatids and the double nature of each chromosome is restored.

Studies employing radioactive isotopes as tracers have added some important information. It is not the chromatids which replicate. It is half-chromatids, also termed chromonemata, which undergo self-duplication. Taylor (1957) placed bean seedlings in a solution containing radioactive (tritiated) thymidine (a substance which the cell uses to synthesize DNA). He left them in this solution until those cells which were in early interphase had each replicated once. He then cut the roots off and transferred them to a nonradioactive solution. In this they replicated a second time. Then, having killed the cells, he determined by autoradiography whether or not the chromosomes were radioactive. Now, if whole chromatids had replicated when they were in the first or radioactive thymidine solution, they would have made for themselves radioactive mates. After cell division half of the daughter chromosomes would have been radioactive, and half, namely, the original chromatids, would have been nonradioactive. But Taylor found that *all* the daughter chromosomes were radioactive! Why? The answer is that half-chromatids replicated, so that each chromatid now consisted of a new radioactive half-chromatid and an original nonradioactive half-chromatid.

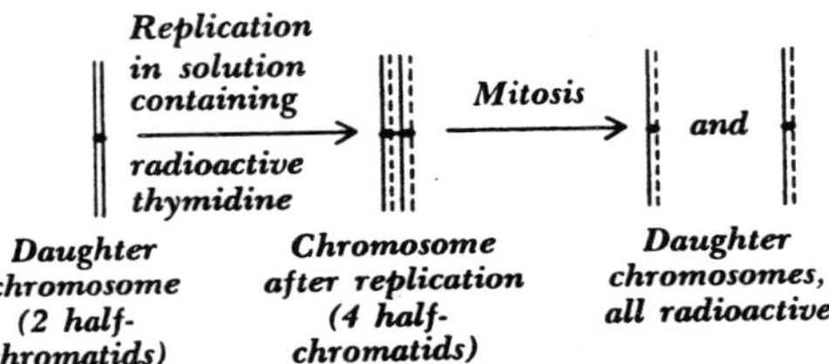

At the second replication, this time in a nonradioactive solution, each half-chromatid, whether radioactive or nonradioactive, made for itself a nonradioactive mate. Therefore, following the next cell division, the daughter chromosomes (chromatids) were half radioactive and half nonradioactive.

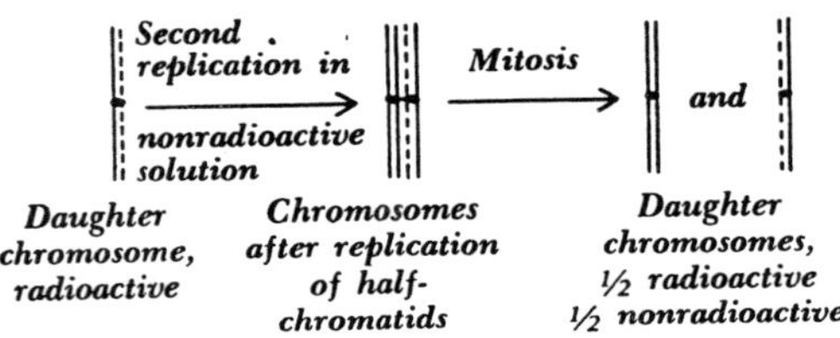

Similar results were obtained by Prescott and Bender, using cells of hamster tissues raised in tissue culture and also human leucocytes.

Experiments of this sort emphasize the remarkable stability of the half-chromatids. Although most of the macromolecules within the cell are in a state of flux, continually breaking down and being replaced, the half-chromatids are passed on intact from one cell generation to the next. And this has been going on for countless generations!

There is evidence that the centrioles may self-duplicate even earlier than the chromosomes, in some cases as early as the preceding anaphase. At this time each centriole (there are two at each pole) makes for itself a partner, not usually by longitudinal splitting, however, as might be expected, but apparently by a process of lateral budding.

The introduction of mitosis in the history of life on the earth was one of the grand innovations of all time. Possibly it took place only once, for, with variations, it occurs in all plants and animals except bacteria and blue-green algae. Even in these there must be a process akin to mitosis, for without some such process the transmission of hereditary characters to each new generation in a balanced fashion would not be possible.

Meiosis

In 1884, Van Beneden described the processes of maturation, fertilization, and cleavage as they take place in the parasitic round worm of the horse, *Ascaris megalocephala.* He demons-trated that the egg and sperm contribute an equal number of chromosomes to the offspring. This led, a few years later, to the discovery of *meiosis,* the process by which the number of chromosomes is reduced to onehalf when the germ cells ripen.

Each body cell and each unripe germ cell or *gonium* has two sets of chromosomes, the *diploid number.* Each of the chromosomes has a mate (except in the case of the unpaired sex chromosomes). When meiosis takes place, an oogonium or spermatogonium, as the case may be, gives away one set of its two sets of chromosomes; that is, it gives away one chromosome of each pair of chromosomes. The result is that each ripe egg or sperm has left only one set of chromosomes, the *haploid number.*

Now, meiosis, unlike mitosis (1) takes place only in the gonads, (2) occurs only in those unripe germ cells which at the time are in the process of ripening, and (3) although it begins earlier, reaches completion only during the period of sexual maturity of the plant or animal. In most cases some unripe germ cells (residual gonia) remain in the gonad where they multiply by mitosis and so produce more gonia. In female birds and mammals, however, the period of multiplication of oogonia comes to an end about the time of hatching or birth. At this time several million gonia may be present; yet only a very few ever ripen into ova—about 400 in the case of women.

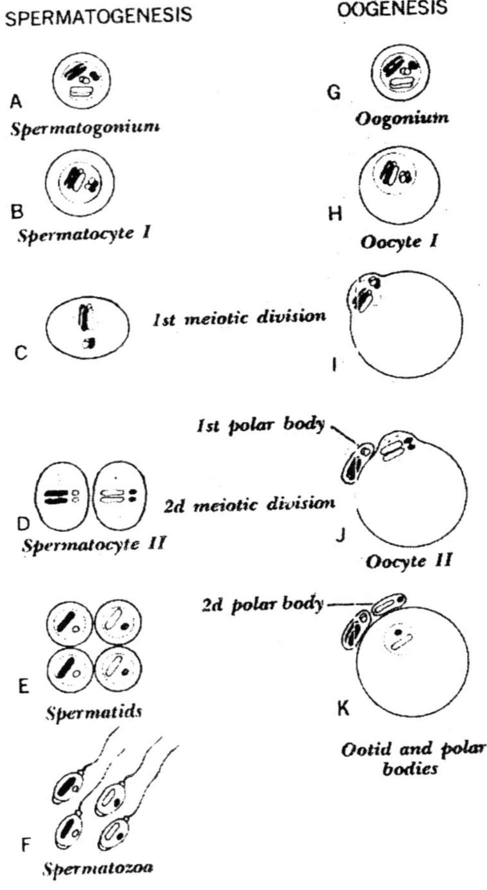

Figure 3.4 : Spermatogenesis and oogenesis. The black ovals represent paternal chromosomes (those derived originally from the sperm). The white ovals stand for maternal chromosomes (derived from the original egg).

The gonia are thus lineal descendants, by mitosis, of the original fertilized egg or zygote. Hence they are cousins of the body cells, and, generally speaking, they possess the same two sets of chromosomes which the body cells possess.

Meiosis has been compared to two mitotic divisions, for two successive spindles are formed, and two separations of chromosomes take place. Yet there is only one complete prophase, and this differs from a typical prophase in that it involves a pairing of the chromosomes side by side, each with its mate. The process of pairing is known as *synapsis*. In a sense this pairing of homologous chromosomes completes the coming-together of the germ cells at fertilization; for all through the mitotic cell divisions of development the chromosomes of the egg and those which came from the sperm have remained separate. Now each chromosome joins its mate.

The stages of the prophase of meiosis have been given names: leptotene, when the chromosomes become threadlike; synaptene (or zygotene), when they come together in pairs; pachytene, when they contract and become tightly coiled; diplotene, when they begin to separate, except for locations where "crossingover" between synapting chromatids has taken place; and diakinesis, when again they become compacted prior to separating.

Note that at the beginning of meiosis each chromosome consists of two chromatids as in ordinary mitosis, so that, when pairing takes place, the result is bundles of four chromatids . These are known as *bivalents* because each consists of two chromosomes. They are also known as tetrads because each is four chromatids.

During this period preparatory to the first meiotic division, the ripening germ cell is called *a primary gametocyte,* primary spermatocyte or oocyte, as the case may be. The diplotene stage of oocytes is of particular interest, not only because it is long drawn out (it may be as long as 40 years in the human female), but because it is a period of great growth. The nucleus enlarges greatly due to the accumulation of nuclear sap and is often referred to as the *germinal vesicle.* The chromosomes become *fuzzy* objects which have been compared to lampbrushes.

First meiotic division

The stages of the prophase which have just been described end as the metaphase approaches. The bivalents move to the equator of the first meiotic spindle. At the metaphase they split; and at the anaphase each bivalent divides, and *a monovalent chromosome* or dyad (pair of chromatids) goes to each pole of the spindle. The monovalents are actually the original chromosomes which united in synapsis, except that some crossing-over has usually taken place. The products of the first meiotic division are known as *secondary gametocytes.*

Second meiotic division

Usually the telophase of the first meiotic division is brief and is quickly followed by the much reduced prophase of the second meiotic division. In some instances, the anaphase of the first meiosis leads directly into the metaphase of the second meiosis. In any case, the monovalent chromosomes (dyads, two chromatids each) take a position at the equator of the second metaphase spindle. At the anaphase the chromatids or *daughter chromosomes* (monads) move apart to the poles of the spindle. The resulting cells are known as *tids,* ootids or spermatids, as the case may be.

Note that a reduction in the number of chromosomes has taken place. The gonia had two sets (diploid number) of chromosomes. The primary gametocyte had one set (haploid number) of bivalent chromosomes (tetrads). The secondary gametocytes had one set of monovalent chromosomes (dyads). Now the tids have one set of daughter chromosomes, that is, one set of chromatids (monads). But this is important: the set which each tid has is a complete set with one chromosome present to represent each pair of chromosomes of the original diploid set.

In the case of spermatogenesis, each primary spermatocyte divides twice equally and produces four equivalent spermatids. In oogenesis, however, although the meiotic divisions of the nucleus are equal, those of the cytoplasm are grossly unequal. At each division almost all the cytoplasm goes to only one of the daughter cells and thus is conserved to supply the substance of the embryo. The result is that the first meiotic division gives rise to one secondary oocyte and one small *first polar body.* The second meiotic division similarly produces one ootid and a small *second polar body.* The spermatids undergo metamorphosis and become sperm cells or *spermatozoa.* The ootids, on the other hand, as a rule need only to burst from the ovary to cause them to ripen and become ready for fertilization.

It is a strange fact that the stage in meiosis which is attained before the fertilizing sperm enters the egg is not the same for all species. In sea urchin eggs meiosis is complete, and both polar bodies have been formed before the egg is receptive to the sperm. This is rather rare. In vertebrates the first polar body has been given off and the second meiotic division has progressed to the metaphase before fertilization takes place. The extreme case is the parasitic roundworm, *Ascaris,* in which the sperm enters the cytoplasm of the egg before even the first meiotic spindle has formed. It remains inactive in the center of the cytoplasm while the egg completes meiosis. In most animal species, therefore, the meiosis which closes the nuclear cycle of the egg overlaps to a greater or lesser extent the entrance of the sperm which begins the nuclear cycle of the new individual.

Why does this complicated process of meiosis exist? The answer was pointed out in 1903 by Walter Sutton, who showed the parallelism between the story of hereditary factors (later termed genes), as worked out by Mendel and other students of plant and animal breeding, and the behavior of chromosomes in fertilization, mitosis, and meiosis. Meiosis, Sutton showed, is nature's way of reducing the number of chromosomes in each germ cell so that, at fertilization, the normal diploid number of chromosomes will be restored. But it is more than this: It is nature's way of seeing to it that the germ cells are all different. Let us assume that the chromosomes of different pairs are different and also that each chromosome has some mutant genes which make it unlike its mate. In nature this is probably always true. Since all the cell divisions during development are made by mitosis, every gonium will have the same two sets, or diploid number, of chromosomes and genes. Then, when meiosis takes place, each ripe germ cell retains only one chromosome of each pair of chromosomes, one gene of each pair of genes. Moreover, it is a matter of chance which chromosome of a given pair of chromosomes the cell retains.

Thus, from the standpoint of heredity, there can always be as

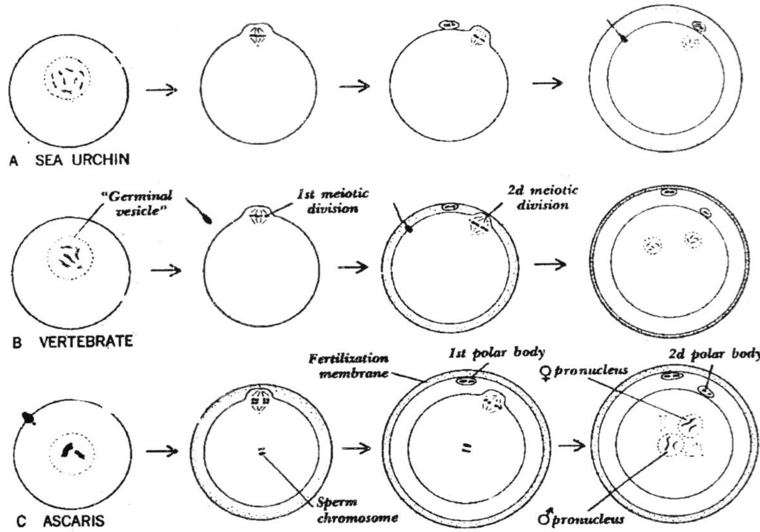

Figure 3.5 : Fertilization (entrance of sperm) takes place at different stages of meiosis in different animals: A, after meiosis is complete in the sea urchin; 8, midway in the second meiotic division in most vertebrates; and C, before meiosis in the parasitic roundworm Ascaris.

many different genetic kinds, or *genotypes,* of ripe germ cells as there are possible combinations of chromosomes taking one chromosome from each mating pair of chromosomes. If there were just one pair of chromosomes in an unripe germ cell, there would be two kinds of ripe germ cells. If there were two pairs to begin with, there would be four possible kinds of germ cells. If there were three pairs, the number would be eight. In general, if there are N pairs of chromosomes before meiosis, there will be 2N possible kinds of genotypes among the ripe germ cells. Now, man has 23 pairs of chromosomes. 2^{23} is 8,388,608. In other words, there is no more than one chance in over eight million that a man or woman will produce two germ cells with identical genotypes. (Crossing-over decreases the chance tremendously.) What is the maximum chance that two identical offspring will be born to the same parents by ordinary sexual reproduction? It is one chance in 8,388,608 times one chance in 8,388,608, or one chance in 70,368,744,177,664. Crossing over at synapsis, mutations, and chromosomal aberrations greatly increase this variability.

THE ROLE OF THE NUCLEUS

The account of fertilization, mitosis, and meiosis which has just been given is similar to that usually found in textbooks of biology. It stops short, however, of explaining how the nucleus controls development. Indeed, it avoids the problem, for the compacted chromosomes of fertilization, mitosis, and meiosis are not active in development. It is only when they are expanded, as at the interphase, that they carry on their functions in growth and development. (The chromosomes of oocytes are expanded also during the long prophase.) Furthermore, the chromosomes presumably are alike in every cell of the body of an embryo. Now, how can chromosomes which are everywhere the same account for the origin of differences within the embryo? In order to approach this problem we must consider the chemistry of the nucleus.

The Chemistry of the Nucleus

The distinctive chemical substances of the nucleus are long chain molecules known as *deoxyribonucleic acid,* or DNA for short. Each such molecule is a gene; more likely it is a string of genes, the material units of heredity. A closely related type of nucleic acid is *ribonucleic acid,* or RNA. Both DNA and RNA are present in the nucleus. DNA is in the chromosomes. RNA is in the chromosomes, but especially in the nucleoli, and it is also present in the cytoplasm.

Now, nucleic acids have three unique and remarkable properties:

1. They are self-duplicating molecules. Possibly they are the only truly self-duplicating molecules. Centrioles and plastids are morphologically self-duplicating, but it is likely that, in this case, nucleic acids are involved.

2. Nucleic acids supervise the synthesis of proteins within the cell. The proteins are the principal structural compounds of protoplasm and are the chemical basis of its functioning. They include the enzymes which control the innumerable chemical reactions of the cell and also structural proteins. In fact, a gene *(cistron)* has been defined as the template for the production of one polypeptide chain. It is probably true that, in the absence of nucleic acids, no proteins are ever produced.

These statements are oversimplified. The self-replication of a nucleic acid requires the presence of specific enzymes (polymerases) and numerous other molecules in the protoplasm. The process is therefore a circular one: nucleic acids bring about the synthesis of proteins, and proteins are needed to make possible the replication of the nucleic acids. It is the entire protoplasmic system which endures.

Genes are not indivisible. Crossing-over of parts of genes may take place between synaptic mates during meiosis. The smallest part of a gene which may interchange thus with a mate has been termed *a recon.*

3. Nucleic acids (genes) occasionally mutate; that is, a short segment of a gene may change in composition. When this occurs, the gene replicates according to its new or mutant form. If this were not so, life on earth would never have evolved, and the ever-increasing adaptation of living things to their environments would not have been possible. The shortest segment of a gene which may mutate is probably a single nucleotide unit. It is known as a *muton.*

Great progress has been made in the last few years in the knowledge of the structure of nucleic acids and of the manner in which they duplicate. According to the Watson-Crick hypothesis, each DNA molecule is a long double chain of units known as nucleotides. It may be compared to a rope ladder twisted into a helix. Each single nucleotide consists of a sugar, deoxyribose (S), a phosphate group (P), and one of four different "bases." The bases are adenine (A, a purine), guanine (G, also a purine), thymine (T, a pyrimidine), and cytosine (C, a pyrimidine). The single chain consists of alternating sugars and phosphates, with a base attached as a side group to each sugar. The precise sequence in which the nucleotides are arranged in the long nucleic acid chain is the basis of the coded information which is passed on by heredity and which guides the course of development.

In the double chain of the DNA molecule, each base of one chain is linked crosswise, by weak hydrogen bonds, with its mate in the other chain. The relations are precise. Adenine is always bonded with thymine, and guanine with cytosine. There are no other combinations. Hence the two chains fit each other exactly like a mold fits a model.

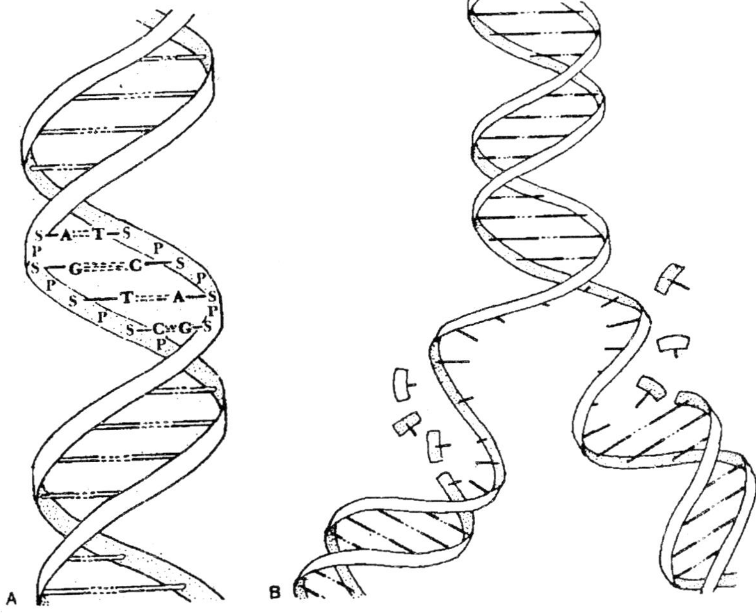

Figure 3.6 : The structure of DNA and the manner of its replication.

When a nucleic acid (DNA) molecule selfduplicates, the chains supposedly separate from each other. Each single chain then makes for itself a replica of its former partner by attracting and assembling to itself appropriate nucleotides from the surrounding protoplasm. The result is that there are now two identical DNA molecules (double chains of nucleotides) where previously there was one. One notes at once the comparison between the replication of the half-chromatids (chromonemata), which takes place during the interphase of mitosis,, and the replication of the single chains of DNA molecules.

The nucleus also contains proteins of two sorts: histones (proteins rich in basic amino acids), which are associated with the genes; and nonhistones, which compose the larger amount of protein and may be a part of the metabolic and mitotic machinery of the cell. The relation between histones and DNA is loose, yet it is so consistent that the combination is usually referred to as nucleoprotein. A small amount of nonhistone protein, termed residual protein, presumably holds the genes in order and organizes them into the definite bodies, the chromosomes. Lipids are also present in the nucleus.

The Influence of the Nucleus on Development

The genes (DNA) do not directly influence development. Instead, they "transcribe" their coded information to RNA, which then collects around the chromosomes and in the nucleoli. Apparently one of the two strands of a DNA molecule assembles ribonucleotides and makes for itself a complementary strand of

RNA. (The other strand possibly self-duplicates.) The nucleotides (ribonucleotides) which compose the RNA, however, differ from those which compose DNA. The sugars are ribose instead of deoxyribose, and one of the four nucleotides is uracil instead of thymine (it lacks the methyl group of thymine). Although these differences from a chemical standpoint appear to be slight, yet they account for important differences in function. The RNA molecules are single chains and as such are not self-replicating. (RNA is self-replicating in some viruses.) They are dependent on DNA for their production. Moreover, the RNA molecules do not remain in the nucleus. From time to time they migrate into the cytoplasm where they function in the manufacture of proteins. Some of them serve as templates for the production of proteins, notably for the production of the enzymes which catalyze the chemical processes of the cell. Possibly this release of RNA is not a continuous process. An impressive release takes place at the prophase of the first meiotic division of oogenesis when the swollen oocyte nucleus (the so-called germinal vesicle) breaks down and its nuclear sap and the material of the large nucleolus mingle with the cytoplasm. A similar release occurs at the prophase each time a cell divides.

In very active cells, such as growing oocytes and certain cells of larval insects, the nuclear membrane has been shown by the electron microscope to be perforated. Nuclear material has been found to "bleb" through pores into the cytoplasm where it supervises the syntheses of proteins.

There may be a delay between the time of the formation of the RNA in the nucleus and the time when its influence becomes apparent in cytoplasmic processes. The clearest evidence that this is so comes from experiments on species hybrids, that is, on eggs which have been fertilized by sperm of another species. In this case the early events of development, for example, the pattern and rate of cleavage, are determined by the egg cytoplasm alone, presumably by the RNA that was synthesized while the oocyte grew in the mother's ovary and under the influence of the mother's genes. Generally speaking, the effect of the genes (DNA) brought in by the sperm is not apparent until about the time of gastrulation. But from then on, the genes of both the egg and the sperm control the course of development.

Even more crucial are certain experiments in which eggs were enucleated before they were .fertilized by the sperm of another species, or in which the egg nuclei were destroyed immediately after the foreign sperm had entered. In such haploid eggs (termed androgenetic hybrids), the cytoplasm belongs to one species and the nucleus to another. Such combinations have not lived beyond the early embryonic stages, but the evidence indicates that, from the time of gastrulation onward, the influence of the chromosomes of the stranger nucleus is pronounced.

Do Nuclei Undergo Differentiation?

In the early days, Weismann and Roux believed that the nuclei of body cells become different during cleavage as a result of what were thought to be unequal divisions of the nuclei. The organization of the embryo, so they theorized, is the expression of the inherited organization within the chromosomes of the zygote nucleus.

This view has long since been given up. There is organization within the cytoplasm of the uncleaved egg. There is strong evidence that the nuclei of every cell of the body contain initially the same set of chromosomes and genes, and that, during cleavage at least, mitosis is equal cell division in so far as the chromosomes are concerned. This fact was originally demonstrated in an experiment by Spemann in which he constricted an uncleaved amphibian egg by tightening a hair noose around it. The egg was pinched into two halves, with only a narrow neck of protoplasm between them. One half contained the nucleus; the other half lacked a nucleus. After several cleavages had taken place in the nucleated half, one of the descendant nuclei, migrated across the isthmus of cytoplasm into the non-nucleated half (delayed nucleation). Cleavage followed, and both halves (provided the needed cytoplasm was present) became whole embryos of half size. This proves that a cleavage nucleus can do everything that the original zygotic nucleus can do. It is an exact copy of its original.

The fact that the daughter nuclei are equal during early development has been clinched by some remarkable experiments by Briggs and King. They removed the nucleus of an uncleaved frog's egg, and substituted a nucleus from a cell of a blastula or an early gastrula. With refined techniques, many such operations were performed and were successful. Normal embryos developed. Here again, it is demonstrated beyond question that at this early stage of development when the future fates of the regions of the embryo are being determined, the nuclei, i.e., the genes of the cells of the different tissues, are equivalent. Since the nuclei are alike, it therefore must be the cytoplasms which are different.

Gurdon and others have extended this work and have found that some nuclei of stages as late as swimming larvae, when transplanted to enucleated eggs, are capable of supporting a complete development. Some of the tadpoles even live to metamorphose into frogs and become sexually mature.

This, however, is not the whole story. It is a common observation that the nuclei of cells of different tissues do differ in size and shape. Somehow during development nuclei do change. The cytoplasm, which supplies them with substance, energy, and necessary enzymes, also affects the visible characteristics which distinguish them. It does more than this: it affects their activity; and, at least in some cases, it affects their composition.

The nuclear changes which arise during development are of

three sorts, and they arise in three distinct ways: (1) by the diminution of the chromatin, (2) by stable changes in the chromosomes, and (3) by the epigenetic control of gene activity by the cytoplasm.

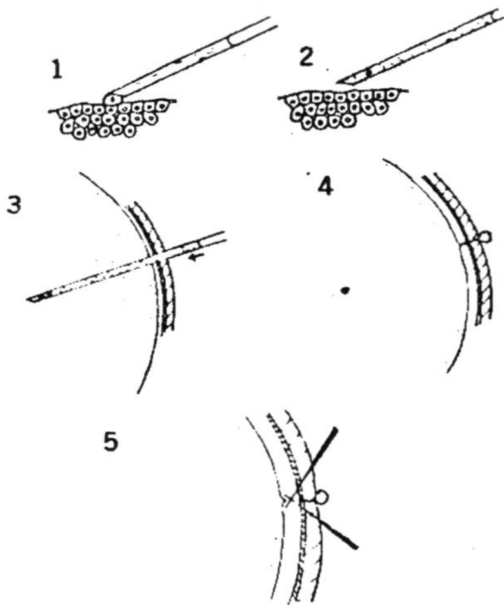

Figure 2.7 : Method used by Briggs and King when they substituted nuclei from older embryos for the nuclei of uncleaved eggs of the frog.

1. Diminution of the Chromatin : This is a rare phenomenon. Boveri observed many years ago that a reduction of the substance of the chromosomes (diminution of the chromatin) takes place during the early cleavages of the parasitic roundworm, *Ascaris*. At the two-, four-, eight-, and 16-cell stages only one of all the blastomeres retains the typical four chromosomes of the zygote. In each case this cell is in the line of descent of the germ cells (germ line). The other cells cast off a part of their chromatin into the cytoplasm; that is, the tips of the chromosomes swell and disappear. At the same time the central part of each original chromosome breaks into several small segments which remain as separate chromosomes.

Other examples of the diminution of the chromatin in the development of body cells have been described, especially among the insects. Sometimes this process takes place early in development, sometimes late. But it never occurs in the germ line. We may conclude that, in the body cells of these embryos, the chromatin (DNA) has already performed its function by producing RNA. The RNA is stable and functions during development. The DNA (chromatin), no longer needed in the development of the body cells, is discarded.

Experiments have shown in *Ascaris* that a certain substance localized in the cytoplasm prevents the diminution of the chromatin. Normally, this material is passed on to only one blastomere during each of the first four cleavage divisions. But if, as a result of manipulation (centrifuging), this material becomes divided between two daughter cells, then diminution of chromatin does not take place in either of them. This is a clear case in which substances in the cytoplasm control the nucleus.

2. Changes in the Chromosomes : Diminution of the chromatin is not known to occur in vertebrate development; yet there is evidence that changes in the genome or sum total of genes may take place. In their studies of transplanted nuclei, Briggs and King found that, when a nucleus from a late gastrula or neurula of an amphibian is substituted for the nucleus of an uncleaved egg, normal development does not usually take place, although it *may* take place. When examined under the microscope, the cells often do not have the normal chromosomal equipment. It is not clear whether changes of this sort take place in normal development. Possibly the nuclei of differentiating cells become increasingly sensitive to ,experimental manipulation. Possibly the transplanted nuclei are forced by the mitotic apparatus to divide before they are fully ready. Possibly, also, when cells differentiate, the mitotic cell divisions become slovenly. In any case, in these experiments the daughter cells did not always receive the normal complement of chromosomes.

The nuclear changes which take place in this manner are stable. Briggs and King have demonstrated that when the nucleus of a defective embryo is transplanted to an uncleaved egg, the result is an embryo with a defect similar to the defect of its nuclear parent. This transfer of nuclei may be repeated over and over with the same result (nuclear cloning).

For a time it was suggested that the changes in the genome which Briggs and King had discovered might be related to the normal differentiation of cells. For example, the nuclei derived from the endoderm seemed to favor the

development of endodermal organs. But this interpretation has not been borne out. Defects of a similar sort have also been obtained when the nuclei are taken from the neural tube. On the other hand, some nuclei are capable of supporting a normal development even when taken from the gut of a swimming tadpole. However, one thing is certain: Any changes which may take place in the genome of differentiating cells always occur *after the* cytoplasm has already become specialized and committed to its fate.

3. The Epigenetic Control of the Genes by the Cytoplasm : Whether or not nuclear substance differentiates during development, it is certain that the same genes are not equally active in every cell and at every place, nor are they equally active at all times. Genes represent potentialities (in the sense of possibilities rather than powers) which may or may not have opportunities to express themselves. Which potentialities are realized depends upon the conditions within the cytoplasm which surrounds a given nucleus; these conditions change epigenetically as development progresses. It has been said that the cytoplasm turns the genes off and on like faucets. To change the metaphor, a nucleus is like a stockroom of a foundry full of patterns. From its vast store, patterns are requisitioned when, where, and as they are needed. No casting can be poured at the foundry for which a pattern is not provided, and, of course, the form of the casting depends upon the shape of the pattern. In the same way, no synthetic chemical process can take place in a cell unless a gene is present to supply the pattern (i.e., template) for the enzyme which mediates the process. Genes are templates in stock. But the decision as to which genes are active at a given time and place-this depends upon cond-itions in the cytoplasm. And these conditions change epigenetically as development goes on. It is a mistake to pass over lightly the epigenetic factors which turn genes on and off, for they, and not the genes, ultimately control the time and place of developmental processes.

Direct evidence that this is true is found in the behavior of the giant chromosomes of certain insect larvae as they approach metamorphosis. Such chromosomes consist of multiple strands of chromatids (so-called polytene structure). They are the result of self-duplication of the chromatids without accompanying separations. For example, the chromatids of the salivary glands of fruit fly larvae multiply nine or ten times without separating, with the result that each giant chromosome consists of 512 to 1,024 chromatids. These compound chromosomes show bands (possibly representing genes) of DNA which stain darkly and are large enough to be seen under the light microscope. They also show swellings, or "puffs," at definite loci. These seem to be the result of RNA and protein accumulating and forcing the strands apart. The puffs may expand until they become what are known as Balbiani rings. Beermann has studied the puffs and rings in the larvae of the midge, *Chironomus,* and has shown clearly that they occur at different locations in the chromosomes of different tissues and at different times. The explanation seems to be that the puffs and rings are centers of intense synthetic activity in particular tissues, in preparation for the role which that tissue will play in the metamorphosis which is soon to follow.

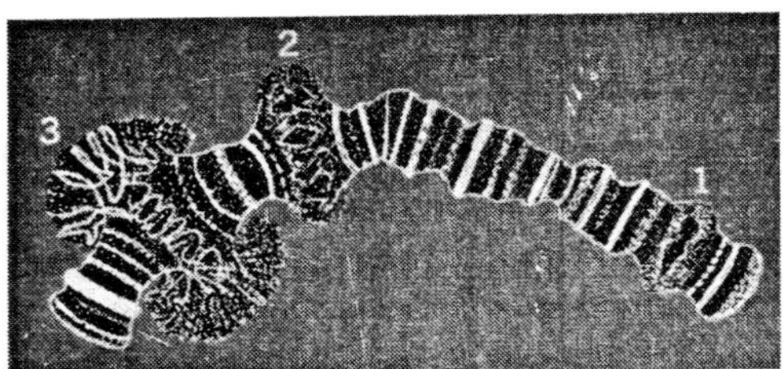

Figure 3.8 : Polytene chromosome of the salivary glands of the midge, Chironomis, showing stages of "puffing."

How are the Genes Controlled?

The analogy of a nucleus to a stockroom of patterns at a foundry has one obvious flaw. A stockroom has a stockkeeper who delivers the patterns as they are called for. The nucleus of a cell, however, is automated so that, cued by conditions in the surrounding cytoplasm, it delivers the templates when they are needed.

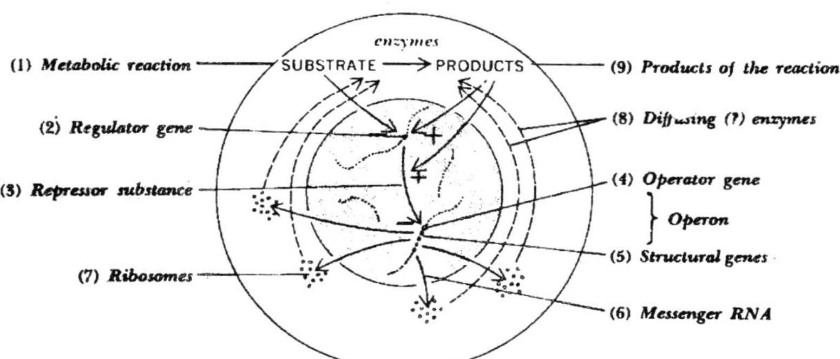

Figure 2.9 : Schema illustrating Jacob and Monod's concept of nucleo-cytoplasmic interaction. 1. A substrate in the cytoplasm inhibits the activity of a regulator gene. 2. A regulator gene, the function of which is to produce a repressor substance. (When inhibited by the substrate it produces less repressor substance.) 3. Repressor substance which prevents the functioning of an operator gene in another chromosome. (When less repressor substance is present, the operator gene is "derepressed.") 4. Operator gene which activates a series of structural genes. 5. Structural genes which synthesize messenger RNA (and other sorts of RNA). 6. Messenger RNA which migrates from the nucleus into the cytoplasm where it associates with ribosomes in the production of specific enzymes (or other proteins). 7. Ribosomes scattered through the cytoplasm. 8. Enzymes which catalyze specific chemical reactions in which the substrate is altered. 9. Products of the reaction, which may augment the activity of the regulator gene or activate a repressor substance. In so doing the products act as inhibitors of the operator gene.

Lately developmental geneticists have been much interested in the problem of how the genes are activated. We shall make reference to the closely reasoned concepts of Jacob and Monod of the Pasteur Institute, who studied enzyme induction in the bacterium, *Escherichia coli.* Normally this microbe produces no enzyme to metabolize the nutrient sugar lactose. Yet, when lactose is supplied in the nutrient medium, the enzyme β-galactosidase, which is able to metabolize it, appears in abundance. The presence of the substrate (substance acted upon) serves as an *"inducer"* and leads to the production of the enzyme.

The authors reasoned that no enzyme could have been produced if a "structural gene" for its production had not already been present in the germinal material. (Genes which determine the structure of a protein are called structural genes.) If this is so, then why was the enzyme β-galactosidase not produced until lactose was added? Genetic studies by Jacob and Monod on mutant strains of *Escherichia coli* supplied them with the answer: The gene for the production of β-galactosidase is inactive because another gene, so-called "regulator gene," is present which produces a *repressor substance* that inhibits the activity of the structural gene. When the substrate lactose is supplied, however, it inactivates the regulator gene (or the repressor substance which it produces). The result is that the structural gene, now freed from repression, gives rise to the enzyme β-galactosidase. The structural gene is thus activated by a process of *derepression*.

This statement is oversimplified. The repressor substance produced by the regulator gene acts upon an "operator gene" within a complex of structural genes. It is the operator gene which directly controls the structural genes, turning them on and off. An operational complex of genes of this sort was called an "operon" by Jacob and Monod. Such a complicated mechanism would be unbelievable if its several steps had not been experimentally verified. Generally speaking, a substrate acts as an inducer. Its presence leads to the production of the enzyme which metabolizes it. In the language of modern communication theory, this is a case of "positive feedback." In a similar manner, the products of a reaction are inhibitors which repress the production of the enzymes which produce them. This is a "negative feedback."

Embryologists are much interested in concepts such as this one of Jacob and Monod because they suggest how epigenetic factors in the cytoplasm (represented in the above case by the substrate lactose) can control the activity of the genes. There is, however, an important difference between the actions of genes in bacteria and in the development of higher organisms. The genes of bacteria are immediately responsive to their environment, and the mRNA which they produce by transcription is immediately active in protein synthesis. It is also short-lived. The genes which control the development of higher organisms, on the other hand, commonly produce "masked" mRNA, which may be stored for a time until the time and place for its activity have arisen.

4

The Cytoplasm in Development

The genes of the nucleus control the course of development, but it is mainly the cytoplasm which develops. The genes are the "physical basis of heredity," but principally it is the cytoplasm which expresses this heredity insofar as it is expressed. There is interaction between the genes and the cytoplasm, but their roles are different. The genes largely control the species characteristics of the new organism and to a great extent its individual characteristics as well; but the characteristics which distinguish one tissue from another, or one organ from another *in the same individual,* are decided initially by processes which take place in the cytoplasm.

The history of the cytoplasm is indeed a true epigenesis; that is, it is a straightforward coming-into-being of molecules, structures, and functions which did not at first exist. It starts in the relative simplicity, or at least plasticity, of a single cell, the unripe egg. It moves forward irreversibly in ever-increasing complexity through the stages of the life cycle: embryo, larva (if any), juvenile, adult, and finally the senescent individual. The ultimate end of development is death unless, as in some of the lower organisms which are capable of regeneration, the story is able to begin again. Embryology is therefore, for the most part, a recital of processes which take place within the cytoplasm.

What is this cytoplasm which, synthesized under the direction of the nucleus, does and becomes so many different things? It is, to be sure, a colloidal aqueous system of proteins, lipids, sugars, and numerous other organic molecules plus various mineral salts. But cytoplasm is more than a chemical system, no matter how complex. It has a structure, an intimate, precise anatomy such as no system of a merely chemical nature ever had. Its proteins and lipids are organized into membranes, filaments, and granules, at the surfaces of which chemical reactions take place. These are the organelles which carry on the many processes of living. The precise patterns of the protein surfaces are the principal basis of the specificity of the cell's chemical reactions, for the proteins include the enzymes which catalyze and control the rates of chemical processes. They also include nonenzyme proteins.

A very few years ago cytoplasm was described as a homogeneous, transparent gel or sol which was optically "empty." The various formed bodies which were seen under the microscope were spoken of as being within the "ground cytoplasm"; or they were interpreted as artifacts produced by fixation and staining. Now this picture has changed. The phase-contrast microscope, the electron microscope, and the ultracentrifuge have revealed an intricate morphologic structure in cytoplasm.

Although the cytoplasms of different kinds of cells have specialized functions (no single type of cell can be said to be "typical"), yet the cytoplasms of most cells posses certain organelles in common. These include: a plasma membrane (cell membrane or plasmalemma, the outer layer of the living cytoplasm); inner membranes, which form vacuoles, tubules, and lamellae; a nuclear membrane (actually a part of the cytoplasm); and formed bodies such as ribosomes and mitochondria. Each of these organelles plays a role in the physiological life processes of the cell. Some of them presumably play a part in development, but it is not always clear what that part may be.

Specialized cells have filaments and granules of various kinds which perform distinctive functions. Muscle cells have contractile fibrils: nerve cells have conductile fibrils; secretory cells have

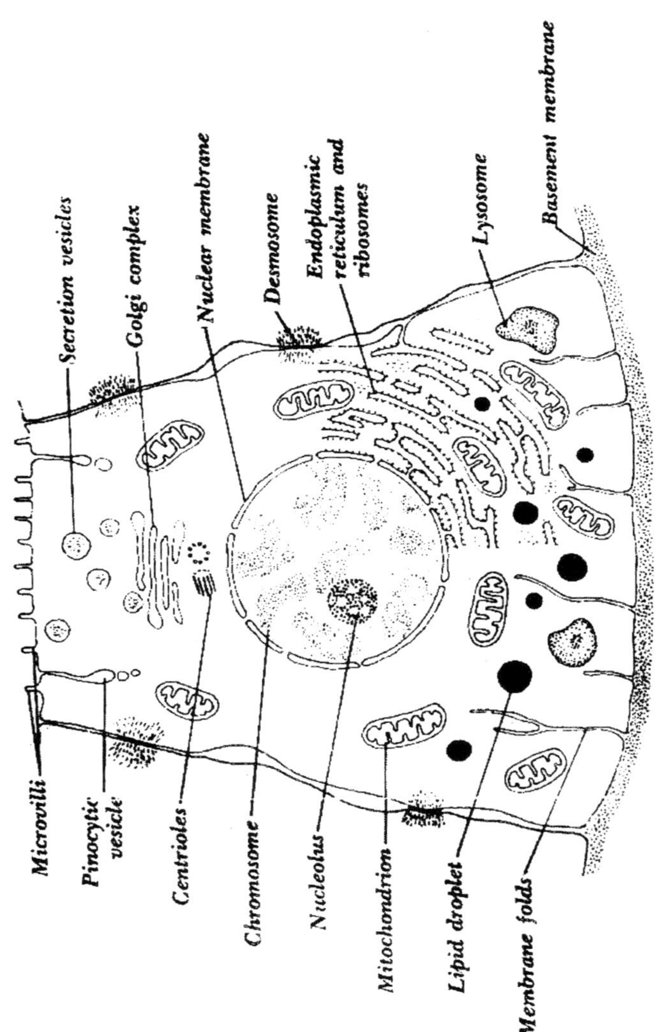

Figure 4.1 : Diagram of an animal cell. The size of the finer structures has been exaggerated.

secretory granules; certain other cells produce pigment granules; while still others store reserve food materials such as yolk platelets, glycogen granules, and fat globules. The list could be multiplied. The embryologist sees these various structures appearing in the cytoplasm as development proceeds, each at the proper time and in the proper place. He has reason to believe that they result from the synthesizing activities of distinctive enzymes which are released in the cytoplasm, and that in turn each enzyme obtains the "information" (specific pattern) for its production from a particular gene or genes in the nucleus. But why a particular gene is active producing a particular enzyme at a particular time and in a particular cell, he cannot as yet say. The answer to this most basic question of embryology is to be sought in the cytoplasm.

Ribosomes and the Synthesis of Proteins

A principal process in development is the synthesis of proteins. Our understanding of the manner in which this is accomplished has been one of the exciting areas of advance in modern biology. Beginning in 1941 and 1950 with the demonstration by Caspersson and Brachet that ribonucleic acid (RNA) is involved

in the synthesis of protein, a great deal of information has been accumulated. As noted in the last chapter, the genes of the nucleus (DNA) "transcribe" their specific genetic "information," coded in the sequence of their nucleotides, to RNA molecules. These accumulate in and around the chromosomes and ultimately enter the cytoplasm. Apparently the DNA produces RNA of three sorts: *ribosomal RNA* (rRNA), *messenger RNA* (mRNA), and *transferor soluble RNA* (tRNA). The ribosomal RNA associates basic proteins to itself and accumulates in nucleoli within the cell nuclei. Later this material (rRNA plus protein) enters the cytoplasm where it appears as ultramicroscopic bodies, the *ribosomes*. As much as 30 percent of the dry weight of cytoplasm may be ribosomes, and about one-half of each ribosome is rRNA.

From time to time messenger RNA (mRNA) also leaves the nucleus and enters the cytoplasm. There it becomes closely associated with the ribosomes and serves as templates, at the surfaces of which the polypeptide chains of amino acids are built up. These chains become the proteins.

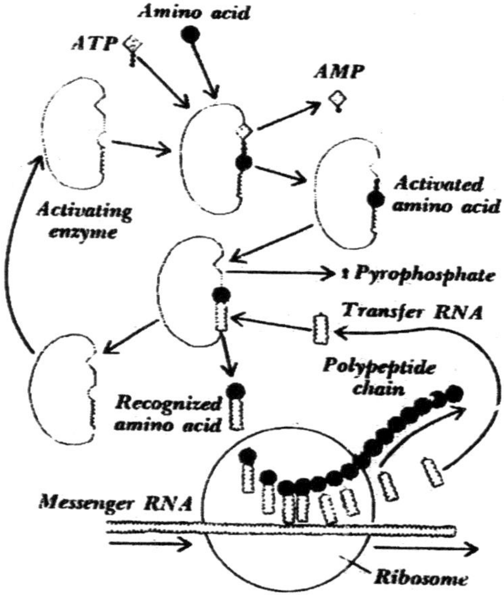

Figure 4.2 : Schema illustrating how synthesis of a protein takes place under the control of messenger RNA.

How is this "information" which is brought from the nuclei by the mRNA transferred ("translated") to protein molecules? It is carried in the form of an alphabetic code of four "letters," the ribonucleotides. These differ from the nucleotides of DNA in that the sugar is ribose, instead of deoxyribose, and the bases are adenine, guanine, cytosine, and uracil (which substitutes for thymine). The precise sequence of the nucleotides is the most important feature, for it determines the sequence of the amino acids in the polypeptide chain. But since there are only four kinds of nucleotides and there are twenty or more kinds of amino acids in a polypeptide chain, it must take a "word" or *codon* of at least three nucleotides to pick or choose one amino acid.

The third sort of RNA is transfer or soluble RNA (tRNA). It consists of smaller molecules, each of which bears, possibly at one end, a triplet of nucleotides which corresponds specifically to one amino acid. Since there are twenty kinds of amino acids, there must be at least twenty specific kinds of tRNA.

According to present evidence, the "translation" of information from mRNA to the polypeptide chain takes place mostly in the cytoplasm in the following manner: (1) First the amino acids in solution in the cytoplasm become activated with the aid of energy-rich molecules, such as ATP, and activation enzymes. (2) Then the activated amino acids, still attached to the enzymes, are "recognized" by specific tRNA molecules. By recognition we mean that each amino acid becomes attached to its proper tRNA molecule. (3) Having accomplished this, the tRNA-amino-acid combinations separate from their enzymes, and each adheres one by one in sequence to a specific locus on a messenger RNA molecule as it moves across and past a ribosome. Thus, in turn, the amino acids, having been brought close together, link up, and a growing polypeptide chain is produced. A polymerizing enzyme is required for the linking-up process.

What is the role of the ribosomes in the synthesis of proteins? It seems not to be specific. Cell-free solutions in which mammalian mRNA have been combined with bacterial (!) ribosomes (plus a "pool" of amino acids, transfer RNA, and ATP) are capable of synthesizing mammalian proteins. One writer has referred to ribosomes as "workbenches" on which the polypeptide chains are forged. Ribosomes are said to "read" the message brought in by mRNA as it moves past. It may help us to visualize the process if we think of a ribosome as a sort of automatic machine which is controlled by a punched tape, the mRNA. The machine (ribosome) receives, as raw material, a mixture of activated and recognized amino acids (amino acids bound to tRNA molecules) and attaches particular combinations as the mRNA tape dictates. The amino acids link up, and an ever-lengthening polypeptide chain is produced. (The tRNA is released to be used again.) When the tape runs out, the polypeptide chain is complete and moves away. The entire process takes but little over one minute. The machine is also free to start over again with the same or another tape. Since the tape is long, it may be moving through or past several ribosomes at the same time, each of them in the process of producing a polypeptide chain. Such a series of ribosomes, connected by a thin thread of mRNA molecules, can be demonstrated with the electron microscope. They are known as *polyribosomes* (polysomes).

(There is some evidence that ribosomes are not all the same in their nucleotide composition, even that they differ in the cells of different tissues. Still, they may all be just workbenches.)

The "code" by which mRNA transfers its information to polypeptide chains has been broken. This was accomplished first by students of microorganisms. Bacteria were homogenized without destroying the ribosomes. The resulting cell-free system was supplied with amino acids, a source of energy (ATP), and artificially synthesized RNA chains of known composition. The polypeptide chains which were produced were then analyzed. The breakthrough came in 1961 when Nirenberg "fed" his system an artificial RNA consisting of the nucleotide urocil only. He obtained polypeptide chains composed of the amino acid phenylalanine. From that point on it was only a matter of time until the entire code was broken.

At first the ribosomes in the cytoplasm are free and scattered. Then, about the time that newly synthesized proteins begin to be secreted away from the cell, the ribosomes become attached to the endoplasmic reticulum within the cytoplasm.

This work on RNA and the ribosomes is of great interest to embryologists because it is possible that differentiation manifests itself first quantitatively in the distribution of the ribosomes. At any rate, gradients in the arrangement of ribosomes exist early in development, and any manipulation which disturbs their distribution, such as centrifuging, results in abnormalities.

Mitochondria and the Energy of the Cell

Whence comes the energy which makes these reactions possible? Proteins are highly ordered molecules, and their synthesis is an uphill (endergonic) process which requires a continuous source of free energy to make it go. If free energy were not provided, the living system would quickly revert to a more probable, that is, a more mixed-up state, namely, death.

Now, the development of an organism is, beyond anything else in the known universe, an improbable sequence of events. Only free energy, that is, ordered energy, can cause it to progress. Furthermore, an organism, and especially a developing organism, is a "dynamic system" like a flame or cyclone. Within it are gradients of substance and energy which would immediately flatten out and disappear if it were not that a continuous expenditure of free energy sustains them.

Where shall the needed order and energy be found? Heat cannot provide it, for it is only when heat is guided from a hotter source to a reservoir at a lower temperature, as in a heat engine, that some of it can be made to do work. Even then, a heat engine is inefficient. Since temperature differences do not exist in the cell, the cell cannot be a heat engine.

Green plants possess remarkable machinery, the chloroplasts, which capture the free energy of light and store it in the form of highly ordered organic molecules, primarily sugars. These molecules, built initially by green plants, are the ultimate source of the free energy which an animal uses in its development.

If the free energy of glucose (about 690,000 calories per gram-molecule) were to be released all in one step, as when sugar burns, most of the energy would become heat, and would be lost. But in the living cell, the energy is released stepwise in a complex series of controlled chemical reactions. In this way, a large part of the energy is transferred to the high-energy phosphate bonds of a ubiquitous substance known as adenosine triphosphate or ATP. Adenosine diphosphate (ADP) plus energy (approximately 12,000 calories per mole) becomes adenosine triphosphate (ATP). The energy-rich ATP can then migrate to where the energy is needed and, like a charged battery, give off its energy in the form of high-energy phosphate bonds, and so revert to ADP.

Perhaps the thing which, more than anything else, distinguishes living from nonliving matter, is this ability to transform and transfer energy in a controlled and useful manner. It is accomplished with the aid of enzymes, or rather, organized brigades of enzymes. One enzyme controls each step of the complex process.

Three successive chains of reactions are involved:

Chain 1 The glycolytic sequence (cf. fermentation)

Glycolysis consists of a series of 14 chemical reactions by which

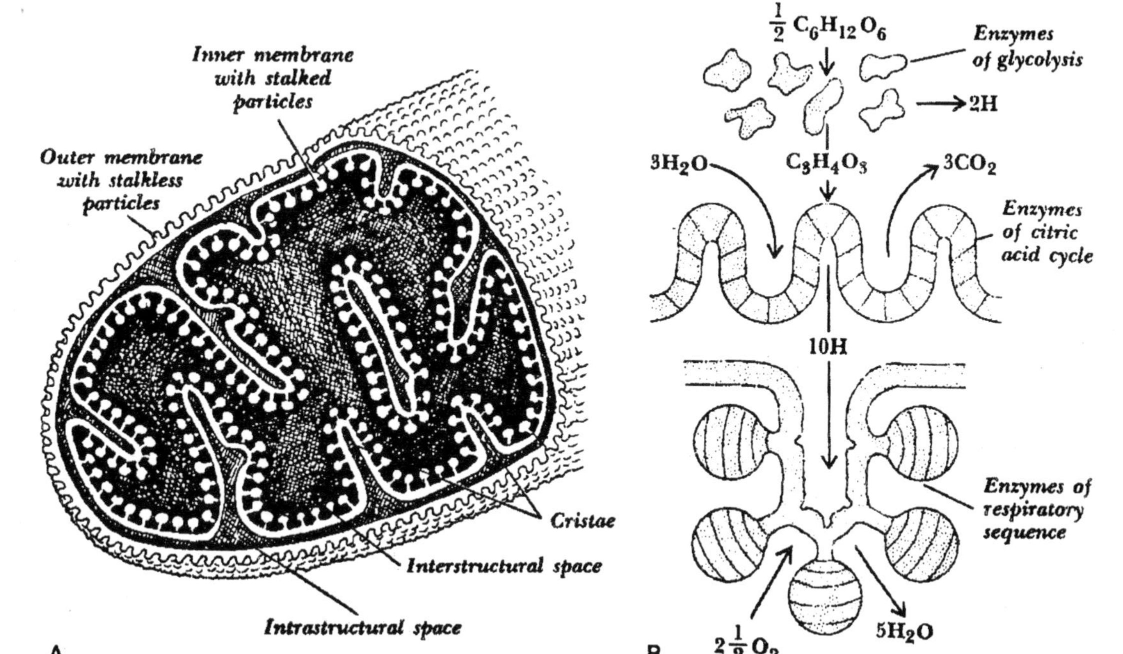

Figure 4.3 : The mitochondrion as a converter of energy. A. The finer structure of a mitochondrion. B. Detail of the outer and inner membranes. The enzymes of gly colysis are probably in the cell solution. Those of the citric acid cycle may be in the outer membrane of the mitochondrion. The enzymes and coenzymes of the respiratory sequence probably occupy the inner membrane.

glucose, glycogen, or other carbohydrate, is split and partially oxidized into a simpler chemical molecule, pyruvic acid. It takes a little energy (supplied by ATP) to start the process, much as it takes energy to start a stone rolling down a hillside; but the ATP which is used to start the process is more than restored by the ATP produced in the process. To control the 14 steps of glycolysis, at least 12 different enzymes are involved.

In the case of anaerobic organisms (those which do not use molecular oxygen), this is the end of the story, except that the resulting pyruvic acid is changed either to lactic acid, to acetic acid and CO_2, or to alcohol and CO_2, without much gain or loss of energy. But the process as it occurs in anaerobic organisms is incomplete, and over 90 percent of the energy of the glucose still remains in the pyruvic acid or related molecules.

Chain 2 The citric acid cycle

Aerobic organisms, on the other hand, namely, those which ultilize oxygen (and this includes most developing eggs), continue the metabolic process by breaking down the pyruvic acid into CO_2 and H atoms. This is done in a closed series of re actions known as the citric acid cycle (of Krebs). It has appropriately been called "the metabolic mill." It is as though pyruvic acid were fed into a hopper and CO_2 and H atoms were ground out. Again specific enzymes and involved, and these act in organized manner. Fats and fragments can also be fed into this same metabolic mill. The CO_2 which comes out is waste and must be discarded, but the H atoms are high in energy. Indeed, a single H atom (gram atom) represents some 52,000 calories of potential energy, four times the energy of an energy-rich phosphate bond.

Chain 3 The respiratory sequence

The third chain of reactions is the one by which the H atoms that were produced in the first two chains are oxidized to water. The energy which is released is transferred to ATP. As before, the process does not take place in one step, for, if it did, a large part of the energy would be lost as heat. Instead, it takes place by a series of steps. At each step a special molecule accepts a pair of H atoms (or electrons-a H atom equals a H^+ ion plus an electron) at a higher energy level., It then transfers them to another molecule at a lower energy level. At some of the steps, ATP molecules are synthesized. The final hydrogen acceptor is oxygen, and the end result is water.

The overall consequence of the three chains of chemical reactions is that, for each sugar molecule which is consumed, 6 molecules of CO_2 are given off, 6 molecules of H_2O are produced, and 38 molecules of ATP are restored. These represent 456,000 calories, or 66 percent of the potential energy of the glucose which was consumed. The remaining 34 percent of the energy is lost as heat.

The enzymes and coenzymes of the last two chains of metabolic reactions are organized in the form of certain ubiquitous organelles which are visible with the light microscope and which are known as *mitochondria.* It is their business to oxidize pyruvic acid and restore ATP. This they apparently do after the manner of a bucket brigade at a fire; that is, they pass on pyruvic acid and its products from locus to locus until it is completely oxidized.

Mitochondria are present in abundance in all cells which are capable of utilizing oxygen in the release of energy. They are most concentrated in the more active cells, such as nerve, muscle, and gland cells, and notably in those regions of these cells where metabolic activity is greatest. A single cell may contain a thousand or more of these bodies. They may make up 20 percent of a cell's dry weight.

As pictured by the electron microscope, each mitochondrion consists of a double membrane surrounding a semifluid "matrix". The inside layer of the double membrane is thrown into internal folds or cristae which partially divide the cavity within. The enzymes of the citric acid cycle (chain 2) are thought to be in the outer layer, while those of the respiratory sequence (chain 3) are localized in the inner layer of the membrane of the mitochondrion. Motion-picture microphotography shows the mitochondria to be in incessant motion, especially at the time of cell division.

Lysosomes

Lysosomes are cell organelles which contain powerful digestive enzymes. Each is surrounded by a semipermeable membrane to prevent the enzymes from escaping and digesting the rest of the cell. Christian de Duve, a biochemist, was the first to recognize lysosomes as such. Later they were identified under the microscope. De Duve noted that when he homogenized rat liver with a Waring blender a very severe treatment-enzymes were released which digested the organic molecules of the cell. When, however, he used a gentler method of homogenization, very little digestion took place. The enzymes were present, but they were held within the membranes of the bodies which enclosed them.

Lysosomes are especially important in white blood corpuscles and other phagocytic cells (cells which devour). They constitute many of the granules of the cytoplasm. Cells of this sort take in foreign particles much as an amoeba engulfs a food particle. The process is called *phagocytosis*. The vacuoles containing the food particles and the lysosomes move together, fuse, and their contents mingle. After a short time the vacuole becomes a digestive vacuole filled with the products of digestion. Absorption takes place, and the residual vacuole, now loaded with undigested debris, is ejected by the cell.

When, under certain circumstances, the membranes of the lysosomes become permeable, digestive enzymes are released into the cytoplasm, and the cell dissolves. This is known as *autolysis*. A controlled autolysis is not uncommon during development. For example, at the time of metamorphosis, the tail of a tadpole undergoes autolysis through the activity of enzymes released from lysosomes.

The Endoplasmic Reticulum

The electron microscope reveals the presence of a net of tubules and hollow lamellae within the cytoplasm of many cells; a light microscope is not powerful enough to reveal this. This net is the *endoplasmic reticulum*. The membranes of the net separate two liquid phases: an outer phase (the ground substance of the cytoplasm), which bathes the net; and an inner phase, which is contained within the net. In some cases, perhaps in all, the membrane of the net is continuous with the plasma membrane and with the nuclear membrane. Thus the fluid which is within the net has connection with the fluid outside the cell. Following cell division, the nuclear membrane seems to be restored from the endoplasmic reticulum.

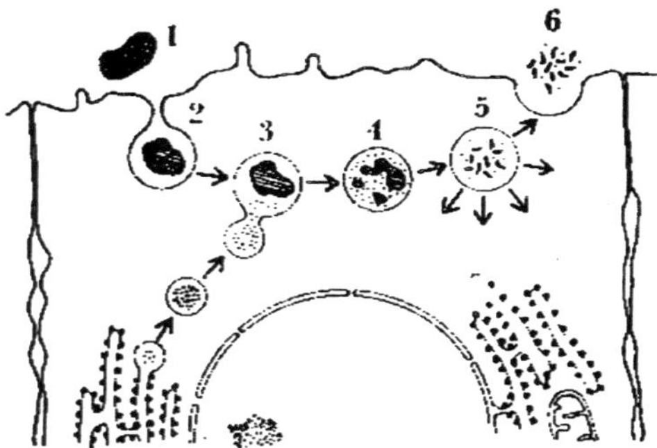

Figure 4.4 : Phagocytosis and intracellular digestion. 1 and 2. A food particle is ingested. 3. It is joined by a lysosome bearing digestive enzymes. 4 and 5. Digestion and absorption take place. 6. Final ejection of waste. The origin of the lysosome from the endoplasmic reticulum is hypothetical.

Endoplasmic reticula are of two sorts: "rough" (granular) and "smooth" (agranular). The roughness of the former is due to numerous ribosomes which adhere to and coat the surface of the net. Because the ribosomes stain with basic dyes, this material of the rough reticulum shows dark with the light microscope. It has been called *ergastoplasm.*

Not all cells show a well-developed endoplasmic reticulum. In early development embryonic cells may possess an abundance of ribosomes, but these are mostly free, and the reticulum is limited in extent and largely smooth. Later, especially in the case of those cells which secrete protein outwardly, the endoplasmic reticulum increases greatly and becomes granular. The implication is that the reticulum is an organelle which receives the proteins that the adherent ribosomes have synthesized, concentrates them, and ultimately delivers them as secretion granules outside the cell. Examples are the gland cells, which secrete zymogen granules, and the fibroblasts, which produce the structural proteins of connective tissue.

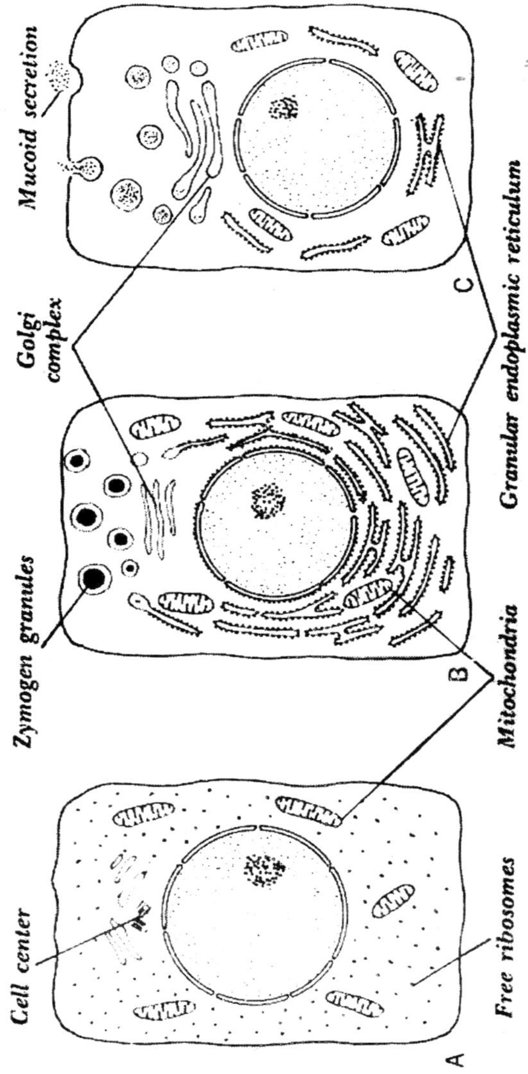

Figure 4.5 : Three types of cells engaged in synthesis (idealized). A. A protein-retaining cell with free ribosomes. B. A protein-secreting cell with an extensive endoplasmic reticulum coated with ribosomes. C. A mucus-secreting cell with a welldeveloped Golgi complex.

The Golgi Complex

Most cells possess in their cytoplasm a region of specialized membranes known as the *Golgi complex* or Golgi apparatus. In thin sections the membranes appear as irregular stacks of flattened vesicles (cisternae) often located not far from the centrioles. The membranes are smooth; that is, they are devoid of a coating of darkly staining ribosomes.

Products of cytoplasmic synthesis accumulate in the Golgi apparatus, possibly having entered from the endoplasmic reticulum. Here condensation and further synthesis takes place, with the result that secretory granules are produced which have carbohydrate (mucopolysaccharides) as well as protein in their composition. The granules collect in vacuoles which, in the case of gland cells, move to the surface of the cell and are discharged to the outside. The process may be thought of as phagocytosis in reverse. Thus the membranes of the Golgi complex act as semipermeable barriers to the free diffusion of molecules, and they may be compared in this respect to the membranes of the contractile vacuoles of protozoa.

The Cell Cortex

The form and activity of a cell are intimately dependent on the properties of the outer zone (or zones) of the cytoplasm, namely, the *cell cortex.* The outer layer of the cortex is the *plasma membrane.* It directly relates the cell to its environment. Nutrients must enter through it, and unwanted substances must be kept out. Wastes are eliminated by it. Stimuli, except perhaps those of light, affect it first. The plasma membrane is of the order of 100 *A* thick (= 100×10^{-7} mm). It has properties of semi permeability, but it must not bethought of as a passive sieve. On the contrary, it is an extremely active membrane which produces and retracts minute processes (microvilli) from its external surface, and forms microtubules and vesicles that push inwardly into the interior of the cytoplasm. By this means the cell feeds (phagocytosis) and drinks (pinocytosis). The particles and droplets which thus enter the cell do not directly penetrate the plasma membrane. Rather, the membrane moves before them and surrounds them as they advance into the cell. Later the membrane may dissolve, or the contents of the vacuole may be digested and absorbed into the cytoplasm.

In addition to the plasma membrane, the cell cortex commonly includes a somewhat thicker layer of cytoplasm which is from 1 to 5 IL thick (= 1 to.5 × 10^{-3} mm). This layer, sometimes referred to as ectoplasm or plasmagel, is usually gel-like in its physical structure, although it may change its state from gel to sol and back again. When photographed with the electron microscope, the plasmagel often differs very little, if at all, from the more liquid internal cytoplasm (endoplasm, plasmasol). Indeed, it may not be present at all. But in other cases it is specialized by the presence of filaments or granules which are not found elsewhere.

The various responses by which a cell changes its shape are often cortical reactions. Development, also, is to a remarkable degree a matter of cortical response. This is obvious in the case of the extrusion of the polar bodies, the rising-up of the fertilization membrane, and the division of the cytoplasm during cleavage. It is not so obvious that the folding of the cellular layers by which the gut and later the nervous system are formed are also to a large extent cortical reactions.

Protozoologists long ago became aware that the form and much of the activity of one-celled animals are properties of the cell cortex. Consider, for example, the ciliate, *Stentor*. The body of this animal is conical in shape. The sides are marked by a hundred or so longitudinal rows of cilia separated by intervening pigmented stripes. The base of the cone (anterior surface or "head") is a disc with spiral rows of cilia. The head is surrounded by an almost complete circle of membranelles. At the clockwise end of the arc of membranelles are the structures used in feeding, namely, the oral pouch, cytostome, and gullet. Nearby is the anal pore. The apex of the cone (posterior end) is specialized as an organ of temporary attachment, the holdfast. All these structures are specializations of the cell cortex.

Tartar has performed numerous surgical operations on *Stentor* and has repeatedly observed its extensive capacity for regeneration

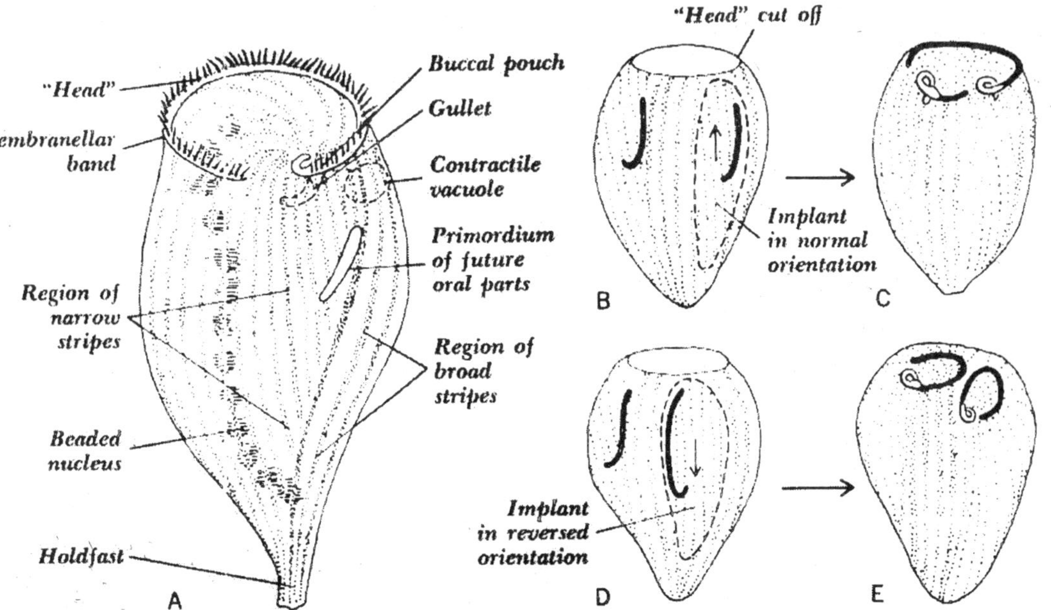

Figure 4.6 : A. Schematic diagram of the ciliate protozoan Stentor. B and C. Tartar's experiment in which a region of the cortex with narrow stripes was implanted in a region of broad stripes in normal orientation. Result-a "doublet" Stentor. D and E. A similar experiment but with the implant revolved 180°. The implant developed a set of oral parts with reversed asymmetry.

and reorganization. The controlling feature, he finds, is the presence and orientation of the cortex. A tiny bit of *Stentor,* only 1/123 the size of a normal animal, is capable of regenerating a whole *Stentor* provided that a little of the cortex and some nuclear material are present. But a far larger piece of *Stentor,* lacking a bit of cortex, does not regenerate even though adequate nuclear material is included. Undoubtedly the nucleus is necessary to control the processes by which substances of the cell are synthesized; but, Tartar concludes, the polarity and asymmetry of the body are transmitted from one generation to the next solely by the cortex.

In a somewhat similar fashion, the cortex of a metazoan egg transmits the pattern which largely controls the polarity, symmetry, and asymmetry of the future embryo. For a time these features are labile to varying degrees and subject to modification by the environment. Raven has studied the role of the cortex in the eggs of mollusks. When such eggs are centrifuged, the visible materials of the cytoplasm are displaced either toward or away from the axis of rotation. They are displaced toward the axis if they are less dense than the fluid cytoplasm which surrounds them, and away from it if they are more dense. Development proceeds normally except that the granules and droplets remain unnaturally distributed. In other cases, the displaced visible stuffs move back to their original locations. Now, since the internal cytoplasm is quite fluid and subject to flow (witness its movements during meiosis, fertilization, and cleavage), it must be the cortex which remains fixed and so transmits the pattern of the developing organism.

(Extreme centrifugation liquefies the cortex and results in abnormalities.)

The Adhesion of Cells

The cells of an embryo adhere to one another in varying degrees and in manners which are characteristic of the different tissues. At first the adhesions are weak. Cells easily dissociate when early embryos (blastulas and gastrulas) are placed in calcium-free solutions. As development proceeds, however, the adhesions between most cells become stronger. This is especially true in the case of epithelial cells, that is, in the case of cells which form layers

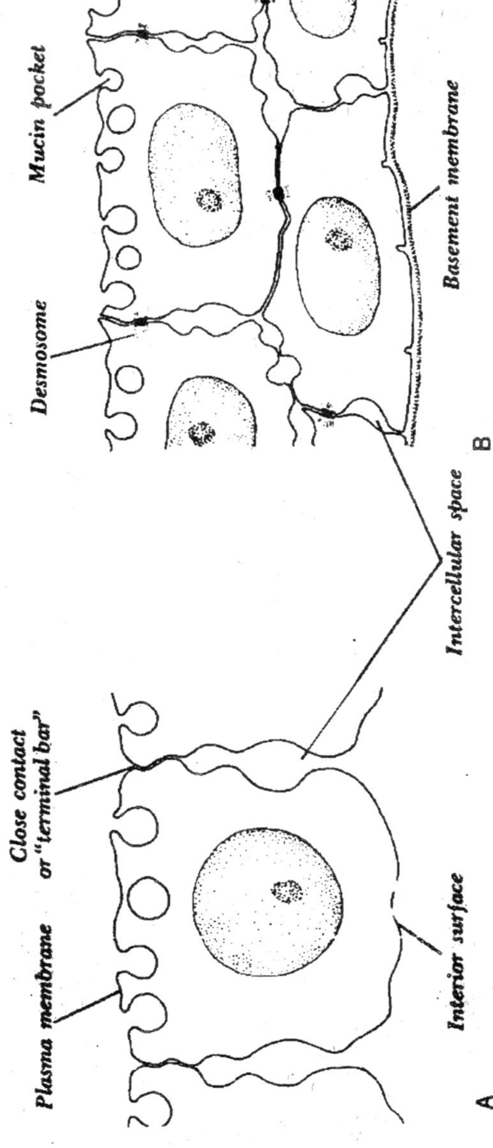

Figure 4.7 : Adhesion of epithelial cells. A. Early one-layered epithelium in which there is close contact next to the external surface. B. Later two-layered epithelium with desmosomes and a basement membrane.

and either bound the outside of the embryo or line cavities within it. Adhesion is especially, close and strong immediately adjacent to where the contacts between cells border a cavity. In effect, an epithelium presents a continuous outer surface-Holtfreter thought of it as a "coat"—which is semipermeable and which serves to separate the external medium which surrounds an embryo (or fills its cavities) from the internal medium which bathes its inner cells. Actually, it is not a continuous coat, but a network of close contacts between cells (often referred to as *terminal bars).* Except at these outer contacts pockets of liquid may separate the plasma membranes.

As development proceeds, adherent patches known as *desmosomes* develop between adjacent cells and strengthen the bonds between them. The electron microscope sometimes shows filaments extending from each desmosome into the cytoplasms on both sides.

Other cells of the embryo, notably those known as mesenchyme, are less adherent to each other. Apparently they are free to wander independently; but they do not wander haphazardly. On the contrary, they move to appropriate positions in the embryo, and there they remain as if trapped by their surroundings.

What guides mesenchyme cells? What finally brings their wanderings to an end? In general, what holds embryonic cells to their places in the embryo?

Holtfreter has attempted to account for the direction of cell migration in terms of "contact guidance." Movement on a surface, he supposes, is controlled by the oriented macromolecular structure of the surface. He explains the final location of cells by the principle of "selective adhesion." That is, cells (or rather, cell surfaces) adhere most strongly to other cells for which they have an "affinity" (possibly a comparable surface pattern). When a cell makes contact with a congenial cell, its wanderings cease. Abercrombie terms the behavior "contact inhibition."

These responses of the cell surfaces are markedly specific. In some cases they appear to be species-specific, as when an egg cell reacts to sperm cells of its own species, or when lymphocytes react to foreign proteins (antigens) by producing antibodies. In other cases—and this is especially characteristic of embryos-the specificity of the cell surface seems to be tissue-specific. When an embryo is experimentally dissociated into its constituent cells and then the cells are permitted to reaggregate, they first adhere to each other and then tend to sort out and reform the tissues from which they were taken. Superficially, it seems as though each cell, when it contacts another cell, recognizes whether the other cell is of its own or another kind.

Steinberg offers a less mysterious explanation of contact guidance and selective adhesion. He explains segregation in terms of quantitative differences in the strength by which cell surfaces adhere. The principle is much the same as that which causes oil and water to separate; namely, the mean of the cohesive forces between like molecules (water to water and oil to oil) is greater than the adhesive forces between unlike molecules (water to oil). Similarly, 'intermingled cells sort out because like cell surfaces adhere to each other with more average strength than unlike surfaces adhere. Sorting out, therefore, is not so much a property of cells as individuals, as it is a property of cells in populations.

A corollary of this is that those cells of a mixed population which have the strongest mutually adhesive properties tend to pull together and become encapsuled within an outer mass of cells with less adhesive tendency.

The foldings and moldings of tissues during development (about which we shall have much to say in the following pages) have been interpreted in terms of programmed changes in the adhesive characteristics of cell surfaces. For example, a thickening of an epithelial layer of cells involves an increase in the areas of contact between cells. Presumably this results from an increase in the strength of their cohesiveness. A thinning of an epithelium is associated with a decrease in the areas of contact.

The Role of the Cytoplasm in Development

Many embryologists today are engaged in describing in detail the chemical and cellular changes which take place in an embryo as development proceeds. This is the problem of *cytodiffere-*

ntiation. Notable progress is being made with the aid of sophisticated physical and chemical techniques such as electron microscopy, radioactive tracers, and immunological methods. As a result, today we know something about the sequence of chemical events in the embryo and when and where new cellular structures appear and disappear. But it is one thing to *describe* the physical and chemical changes which take place during development, and quite another thing to give a *causal account* as to how these changes are brought about. The key question is: What controls differentiation?

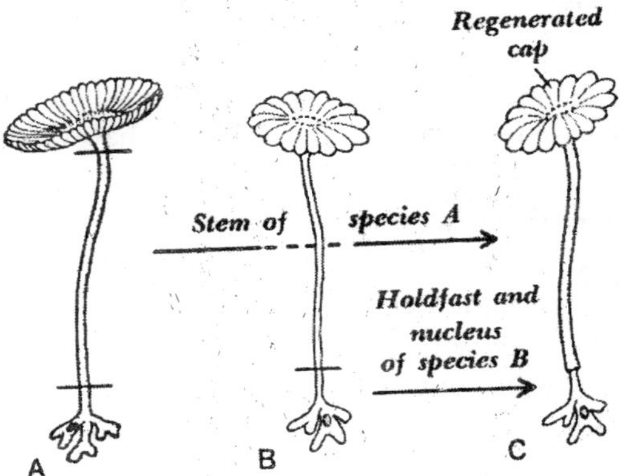

Figure 4.8 : Regeneration in the unicellular alga;* Acetabularia. *A. Acetabularia mediterranea ("species A"). B. Acetabularia crenulata ("species B"). C. Result of grafting a decapitated stem (no nucleus) of species A to a holdfast (with nucleus) of species B. The first cap which regenerated was cut off. The second cap to regenerate had the characteristics of species B.

There is no doubt today that the cytoplasm controls the course of cell differentiation. But there are two "levels" at which this control may be exerted: (1) It has been generally assumed that control takes place at the level of "transcription"; that is, that it takes place in the nucleus when DNA becomes activated and transfers its information to messenger RNA. Some genes are "turned on" at any given time and place, while other genes (probably most of them) remain repressed. (2) There is, however, increasing evidence that some control by the cytoplasm takes place at the level of "translation," namely, when mRNA supervises the synthesis of protein. In this case the mRNA produced in the nucleus remains "masked" until it reaches the proper time and place in the living system.

The classic example of control of differentiation at the level of translation is seen in the marine alga, *Acetabularia.* This is a giant cell, some two inches long, and shaped like a mushroom or parasol. Its apical end spreads out radially like a hat or cap. Its basal end forms rhizoids (holdfasts), one of which contains the large nucleus. The entire life cycle of *Acetabularia* does not concern us now. Our present interest is in its capacity to regenerate a new cap when the old cap is cut off. Surprisingly, the regeneration of a cap in *Acetabularia* takes place even when the nucleus has been cut away two or three weeks previously. Clearly the presence of nuclear DNA is not directly responsible for the restoration of lost parts.

Hammerling, who studied these matters intensively, interpreted his findings in terms of gradients in the distribution of morphogenetic substances, i.e., a cap-producing substance with its highest concentration at the apical pole, and a holdfast-producing substance centered at the basal pole. His concept is that these gradients determine the location of the new parts.

The role which the nucleus plays in regeneration must be an indirect one. Presumably the genes of the nucleus produce the morphogenetic substances which then migrate within the cytoplasm and concentrate at the apex or base, as the case may be. There they stimulate *regeneration and at the same time control the* character of that which regenerates. This interpretation is borne out by Hammerling's observation that, when a piece of the stem of *Acetabularia mediterranea* (we shall call this "species A") is grafted by its basal end to a holdfast of *Acetabularia crenulata* ("species B"), the cap which regenerates resembles the caps of species A. He interprets this to signify that, at the start, regeneration is under the control of the morphogenetic stuffs which were emitted by nucleus A before it was removed. But if this first cap is cut away, the second cap which regenerates has the character of species B. This confirms the view that the nucleus, which is located at a distance from the apex both in time and space, is the source of the morphogenetic stuffs.

In terms of present theory we may identify the morphogenetic stuffs (as Brachet does) *with messenger RNA. They are produced by* transcription in the nucleus. Under the influence of the cytoplasm, the cap-forming stuff migrates and concentrates at the apical end of the cell. Here translation takes place; that is, the information brought by the mRNA is transferred *in situ* to newly synthesized enzymes and structural proteins. No theory of turning genes on and off is sufficient to account for this localized reaction.

What brings about the concentration of morphogenetic stuff at the apex? Since the inner cytoplasm of *Acetabularia* is engaged in active streaming, some factor in the cortex must be responsible. But it must be admitted that, as yet, we know very little of how this influence of the cortex originates or is exerted.

5

Early Development

Viviparity, in which an embryo or fetus is nurtured within its mother's body, is characteristic of mammals. Exceptions are the primitive monotremes of Australia and Tasmania, namely, the platypus, *Ornithorhynchus,* and the spiny anteater, *Echidna. Viviparity* is an adaptation to adverse environmental conditions such as cold, dryness, and seasonal change. It permits the mother freedom to move about, to feed, and to migrate, freedom which is not enjoyed by a bird parent during its task of incubation. Viviparity also is found in some elasmobranchs, teleosts, and reptiles. In most of these, however, the egg is supplied with abundant yolk, but develops within organs of its mother's body until hatching. Birds apparently have not developed vivi parity because of the burden of carrying the unborn young when in flight. However, bats, which are mammals, carry their offspring in flight.

THE OVARIAN FOLLICLE

At the time of birth, the ovary of a typical female mammal contains a hundred thousand or more oogonia (potential eggs). Some are still in the process of growth and mitosis. But soon mitoses cease, and during the remaining life of the female, no further multiplication of germ cells takes place.

At first the oogonia are associated with the mesodermal epithelium (germinal epithelium) which covers the ovary. Whether they are actually derived from the germinal epithelium has been

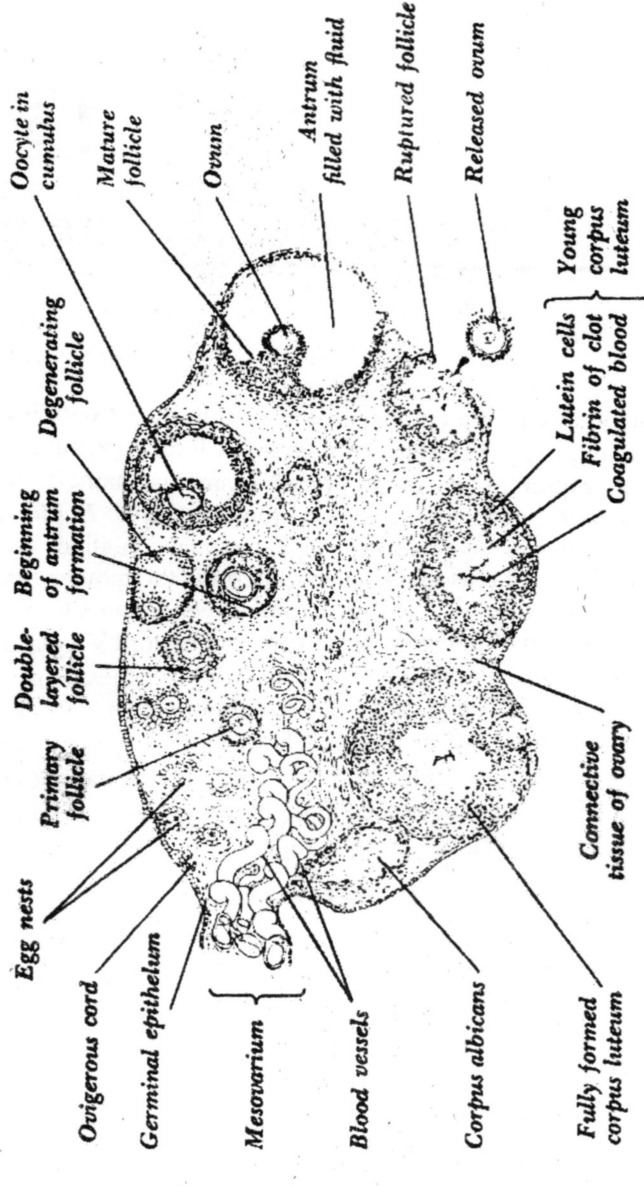

Figure 5.1 : Diagram illustrating the sequence of events in the mammalian ovary.

debated. At any rate, cords of cells from the epithelium, including germ cells, push inward into the connective tissue of the ovary (ovigerous cords). Here they break up into "nests" of cells, each consisting of a central germ cell-now termed an oocyte—and, surrounding it, a single layer of follicle cells. The entire cluster is *a primary ovarian follicle.*

Most of the oogonia do not reach this stage, and of those that do, only a very few ever complete the process and become ripe ova. In the entire life span of a human female, from the time of puberty (about 11 years of age) to the menopause 30 or 40 years later, normally only one ovum matures each month. This makes a total of 400 to 500 ova. Some mammals, on the other hand, release several ova at each ovulation.

The primary ovarian follicle increases greatly in size as a result of the multiplication of the follicle cells and the growth of the oocyte at its center. When its wall becomes several cells thick, fluid-filled vesicles appear in the midst of the follicle cells and begin to coalesce. Finally the mature follicle resembles a blister on the surface of the ovary. It is then known as a *Graafian follicle,* after the seventeenthcentury physiologist who first described it. Such a follicle consists of an outer layer of follicle cells surrounding a central cavity filled with follicular fluid. On one side of the cavity, attached to its wall, is a mass of inner follicle cells (the cumulus) with a maturing ovum at its center.

OVULATION AND FERTILIZATION

At the proper time in the ovarian cycle, the mature follicle bursts. Follicular fluid flows out, sweeping the ovum along with it. The follicle cells which accompany the ovum form what is known as the *corona radiata.*

They soon disappear. Normally, the ovum is swallowed directly into the mouth of the oviduct (uterine or fallopian tube) and is carried by ciliary and muscular action toward the uterus. If sperm are present, they will have been propelled rapidly to the upper end of the oviduct by the contraction of the muscles of the uterus and tube. The capacity of the sperm to swim apparently plays but little part, for dead sperm have been found to arrive at their destination as rapidly as living sperm. Fertilization takes place as the egg enters the tube and begins its journey to the uterus.

Occasionally an egg may fail to enter the oviduct but instead remains in the body cavity. It may even be fertilized within the body cavity. In this case the placenta, or organ of exchange between the embryo and the mother, forms against the peritoneum of the body cavity.

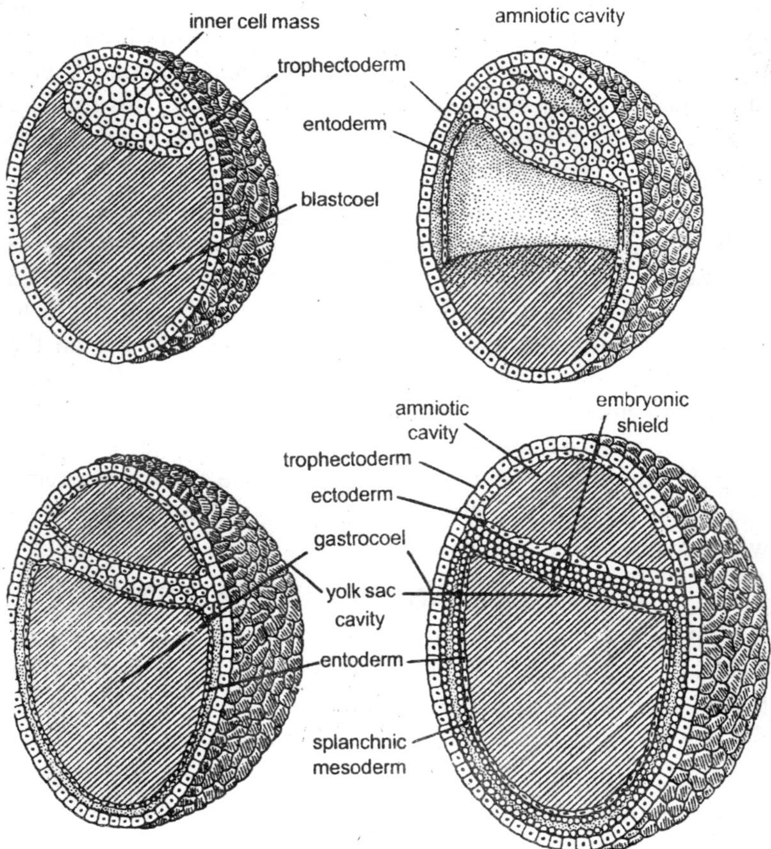

Figure 5.2 : Stages in the development of the mammal egg showing the formation of blastocoel and gastrocoel.

The ability of eggs to find and enter the mouth of an oviduct is remarkable. They may even cross the body cavity from one side to the other. In an experiment on a sow, an ovary was removed on one side, and the oviduct was tied (ligated) on the other side. Not only did several eggs cross the body cavity and enter the oviduct of the contralateral side, but when they reached the sow's two-horned uterus, they spread apart and spaced themselves in both horns.

THE MAMMAL EGG

The mammalian ovum which bursts from the ovary is typically a cell from 70 to 200 μ in diameter. (One micron, or μ, is one-thousandth of a millimeter.) The mouse egg is 70μ; the human egg is twice that size, 140μ. It is a speck which is barely big enough to be seen with the unaided eye. Yet as cells go, it is large. A red blood corpuscle, for example, is 8μ in diameter.

The egg is surrounded by a tough outer membrane, the *zona pellucida* (not to be confused with the area pellucida of the hen's egg). Probably the zona pellucida is a product of both the cytoplasm of the oocyte and the surrounding follicle cells.

A typical mammal egg possesses little, if any, yolk. Hence it must be classed as microlecithal, or isolecithal, since what little yolk it has is evenly distributed. Its nucleus is central. Its cleavage is complete (holoblastic). Yet, in its later development, a mammal egg resembles the development of the yolk-rich cleidoic eggs of reptiles and birds. This can be explained by the proposition that the mammal of today is descended from ancestors which had yolk-rich eggs.

The most primitive mammals which are still living, the monotremes, lay yolk-rich eggs which are surrounded by albumen and a shell secreted in the oviduct. Their cleavage is incomplete (meroblastic) as in the case of the eggs of birds and reptiles. The eggs of some marsupial mammals, the opossum, for example, also receive a secretion of albumen as they descend the oviduct. They even possess a delicate shell membrane. Albumen is also deposited around the eggs of some placental mammals, such as the rabbit. Even in man the oviduct secretes a nutrient fluid which nourishes the egg until it has opportunity to implant itself in the uterus.

CLEAVAGE

Cleavage takes place as the egg moves slowly down the oviduct, driven along by cilia and muscular contraction of the wall of the duct. The journey takes about four days in the case of the human species.

The first cleavage is slightly unequal, and the plane which separates the first two blastomeres does not necessarily pass through the pole where the polar bodies were given off. In the case of the pig, according to Streeter, the smaller of the first two blastomeres gives rise to cells which become the embryo proper. The larger blastomere becomes the extra-embryonic ectoderm (trophoblast).

The significance of these observations has not been experimentally confirmed. Whatever the normal fate of the first two blastomeres, there is good reason to believe that they have like capacities. Nicholas and Hall found that in the rat, if the first two blastomeres are. separated, each can give rise to a whole young embryo. It is even possible to combine two fertilized eggs and obtain a single embryo. Others have noted that the early blastomeres are loose and roll around within the confines of the zona pellucida. Furthermore, a single mammal egg may give rise to two or more embryos. Facts such as these make it uncertain as to whether the first two blastomeres differ morphogenetically from each other. Certainly their fates have not been irreversibly determined.

As a result of early cleavages, the egg becomes a solid ball of cells known as a *morula* (Greek for "little mulberry"). It is not a blastula, for it lacks a blastocoel. Rather it should be compared to the blastodisc of a bird's egg. It is as though the yolk of the bird's egg (during cleavage) had been removed and the cells compacted into a ball.

THE BLASTOCYST

Soon the cells of the morula rearrange themselves and develop a cavity in their midst. The structure is known as a *blastocyst,* or blastodermic vesicle. What is the cavity? In the past it has generally been considered to be a blastocoel. But it is not comparable to the blastocoel of an amphibian blastula, for the latter separates prospective ectoderm of the animal hemisphere from prospective endoderm of the vegetal hemisphere. The cavity of the mammal blastocyst, on the other hand, separates cells which will become the embryo proper-the so-called *embryonic disc* or inner cell mass-from cells which will form no part of the embryo whatever. Instead, the latter cells become a thin outer layer, the *trophoblast,* which soon comes into contact with the wall of the maternal uterus.

The cavity of the mammalian blastocyst is comparable to the yolk region of a chick egg; or more precisely, it corresponds to the subgerminal cavity that separates the blastoderm from the mass of yolk beneath it. The embryonic disc of the mammal egg represents the central region of the chick blastoderm (the area pellucida). The trophoblast corresponds to the ectoderm of the outlying regions of the blastoderm, namely, to the ectoderm of the area opaca.

The size of the embryonic disc varies from species to species. In some mammals, notably, in some marsupials, it is at first not recognizable at all. In others, for example, in man, a knob protrudes into the cavity of the blastocyst hence the name, "inner cell mass." In this case, only the deeper cells of the inner cell mass are actually the embryonic disc. The surface cells which cover the mass are extra-embryonic like the trophoblast.

THE DEVELOPMENT OF THE FETAL MEMBRANES

Great variety exists among the mammals regarding the manner in which the blastocyst gives rise to the embryo and its fetal membranes. Typically, the embryonic disc (inner cell mass) splits (delaminates) into two layers in the same way in which the blastoderm of the chick splits. The underlayer is the *hypoblast.* The overlayer is the *epiblast.* The cells of the hypoblast spread laterally, as in the chick, and finally close to form a complete yolk sac.

In discussing the chick, we noted that two interpretations of delamination are possible: (1) According to the usual view, delamination is a mode of gastrulation. The hypoblast, which splits away from the epiblast, represents the roof and wall of the gut. (2)

According to the other view, delamination is a method of blastulation. The cleft which forms between the epiblast and hypoblast corresponds to the blastocoel of an amphibian egg. If this latter interpretation is correct, the hypoblast represents the floor, rather than the roof, of an amphibian archenteron.

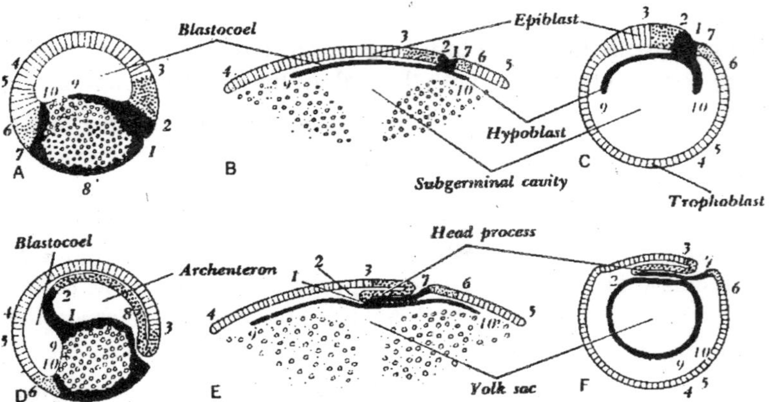

Figure 5.3 : An interpretation of the homologies of blastulas and gastrulas. A and D. Blastula and gastrula of an amphibian. B and E. The primitive streak and head-process stages of a duck or reptile. C and F. Comparable stages in mammal development.

Cell movements of the embryonic disc take place in the same way that they do in the chick. Cells of the epiblast stream toward the posterior quadrant of the disc. Here they heap up as the *primitive streak.* From the undersurface of the streak, mesoderm pushes out laterally between the epiblast and hypoblast. Soon—in some cases at its very beginning-the mesoderm splits into two layers. The outer layer is somatic mesoderm. The inner layer is splanchnic mesoderm. The extra-embryonic coelom lies between the two. The trophoblast and somatic mesoderm together give rise to the chorion and amnion. The splanchnic mesoderm and hypoblast produce the yolk sac and allantois.

The *chorion,* since it is the outer layer, is the place where exchange of substances occurs between embryonic tissue and the maternal environment. To facilitate the exchange, various adaptations have been acquired. In the pig, the chorion elongates tremendously. Its great area affords abundant contact with the uterus wall.

In most mammals, the chorion develops more or less complex outgrowths known as *villi.* These penetrate into the tissue of the uterus wall and greatly increase the area and intimacy of contact between fetal and maternal tissues. Usually there is a specialized region of the chorion where exchange between the mother and embryo is most efficient. Such an area is known as *a placenta.*

The *amnion* may be thought of as a membranous tent around the embryo. It contains a fluid, the *amniotic fluid,* which bathes the outside of the embryo. Its function is to free the embryo from pressures, permit freedom of 'motion, and protect the outer surface of the embryo from abrasion.

In more primitive placentas, such as those of the pig, the amnion arises in the same way it arises in birds; that is, it is formed by the rising up of folds from the extra-embryonic somatopleure. These close in above the embryo. In some of the more advanced mammals, on the other hand, the amnion seems to form in a different and more direct manner. The amniotic cavity makes its appearance as a cleft in the inner cell mass. This method of forming the amnion is known as *cavitation*. This is a sort of shortcut in development which accomplishes the same result as folding. The cells which are above the cavity correspond to the amniotic folds (chorion and amnion). The cells which are below the cleft, that is, between the cleft and the yolk sac, are the embryonic disc proper. They alone give rise to the embryo. In the guinea pig, the amnion first forms by cavitation; then it opens out and exposes the embryonic disc; finally, it closes in again by folding.

In all cases the amniotic cavity expands until it completely surrounds the embryo, except, of course, where the umbilical cord is attached.

The *yolk sac* also varies from species to species. But why should a mammal have a yolk sac at all? Little or no yolk is present in the egg to be digested and absorbed. One answer, no doubt, is an evolutionary one. The ancestors of present-day mammals had yolk sacs full of yolk, and the mammals today recapitulate the developmental history of their forebears.

This, however, is an inadequate answer. The yolk sacs of modern

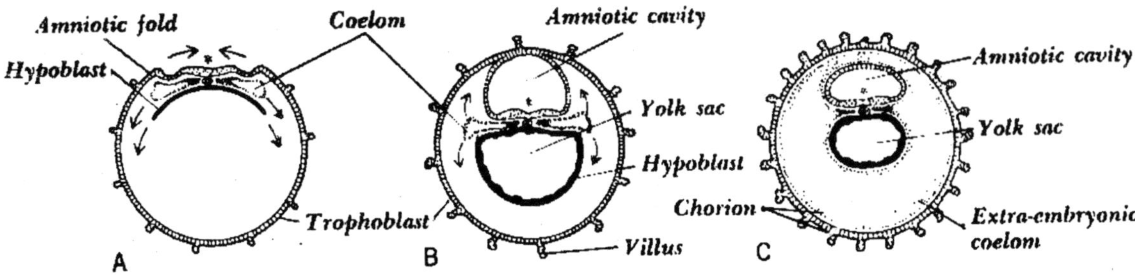

Figure 5.4 : The development of the fetal membranes in mammals. A. In the pig, the amnion is formed by folding and the yolk sac by outgrowth, much as in the chick. B. In most groups of mammals, the amniotic cavity and the yolk sac are formed by cavitation within the inner cell mass. C. In man, even the extra-embryonic coulomb is formed by cavitation. In each figure, the embryo proper (at open neural plate stage) lies just beneath the asterisk.

mammals do have a function. To a varying degree they are organs of exchange of substances between the embryos and the maternal fluids which surround them. Even in the case of the highest mammals, the yolk sac probably carries on some respiratory exchange of gases before the allantois becomes functional. The yolk sac at first is close to maternal tissue. It is separated from the maternal blood by the thin ectoderm and mesoderm of the chorion. In most marsupials, a *yolk-sac placenta* is formed. This is a specialized region where the vitelline blood vessels of the yolk sac carry on exchange of substances with the mother. Even in some placental mammals, the pig for example, the yolk sac, for a time, is large and probably functional. It enfolds the embryo and its amniotic covering. In most other placental mammals-and this includes man-the yolk sac is small and is separated from the chorion by the extra-embryonic coelom.

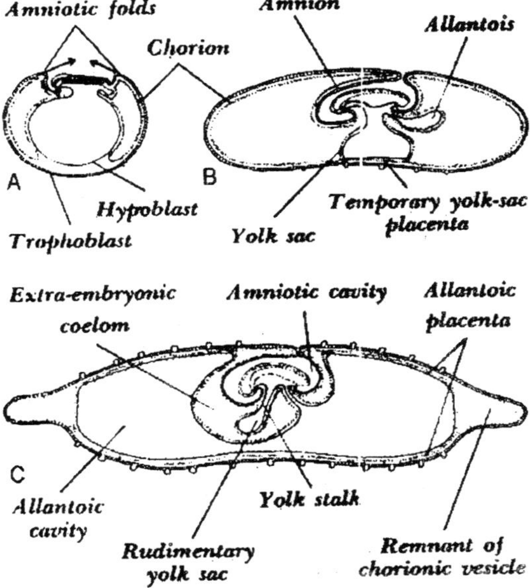

Figure 5.5 : The development of fetal membranes in the pig. Highly diagrammatic. Ectoderm is shown by a laddered line, mesoderm by stippling, and endoderm by a solid line. The cavity of the allantois is lined diagonally. A. Formation of amnion by folding. B. Early stage with a temporary yolk-sac placenta. C. Fully developed allantoic placenta.

Even when the yolk sac is not an organ of exchange with the mother, it has important functions. In mammals, as in birds and reptiles, it is the organ in which the first blood cells are formed. For a time, it probably, also, has certain synthetic and storage functions similar to those which are later performed by the embryonic liver.

Finally, the *allantois* varies greatly in different mammals. When describing the development of the allantois of the chick, we compared it to a precocious urinary bladder which early pushes out from the hindgut of the bird embryo and spreads around the embryo next to the shell membranes. Its mesoderm fuses with that of the chorion. Its rich plexus of umbilical blood vessels acquires the role of transporting respiratory gases.

In placental mammals and in some marsupial mammals, the allantois similarly fuses with the chorion. Together they produce an allantoic placenta for the effective exchange of substance with the maternal environment. In the pig, the allantois expands tremendously. Its cavity serves as a sac in which wastes from the embryo accumulate. In man, on the contrary, the cavity of the allantois is rudimentary. In fact, the human placenta is so efficient that the wastes of the embryo pass into the maternal bloodstream and are excreted by the mother's kidneys. Yet, even though the endodermal part of the human allantois is small, its splanchnic mesoderm is large and develops precociously. In fact, it is present and already fused with the mesoderm of the chorion when the cavities of the amnion and extra-embryonic coelom first appear.

In embryology there are many examples of advanced animals which have adopted shortcuts in their development. The development of the fetal membranes is an instance in point. Primitive vertebrates produce the membranes by the more laborious and circuitous routes of folding. Higher forms have adopted the more direct method of cavitation. Of course, we do not know what morphogenetic movements may have taken place before the different cavities appear. Certainly, if actual morphogenetic movements do not take place now in the development of the individual, they did occur in the past phylogenetic history of the race.

THE DEVELOPMENT OF THE EMBRYO PROPER

The great variety which is shown in the development of the fetal membranes is not duplicated in the development of the embryo proper. The processes of neurulation and embryo formation of all mammal embryos are remarkably alike. In fact, embryo formation in the mammal is so similar to that in the chick.

The embryonic disc of the mammal is comparable to the area pellucida of the unincubated chick egg. Both give rise to the embryo proper. In both, the mass movement of the epiblast is toward the posterior quadrant. In both, also, this movement leads to the formation of the primitive streak. Again, in both, the cells of the primitive streak sink inward and spread as mesoderm between the epiblast and hypoblast.

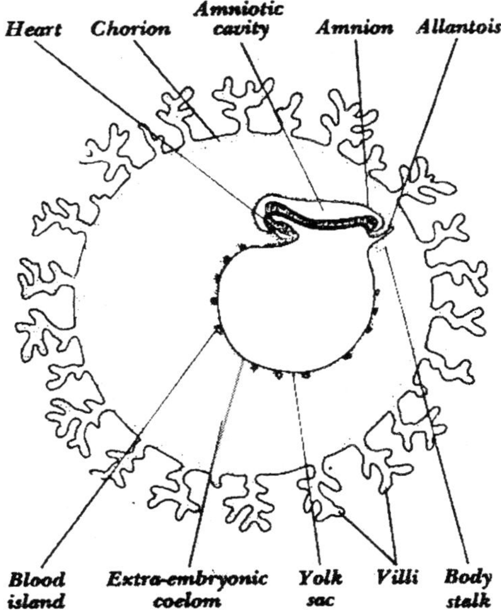

Figure 5.6 : A human embryo surrounded by its membranes (schematic). Note that the allantois (endodermal) is rudimentary although its mesoderm is greatly developed.

Of special importance is the *head process* which pushes forward from the anterior end of the primitive streak. In the mammal, as in the duck and many reptiles (but not in the chick), the head process develops a central canal (notochordal canal) which ends posteriorly at the primitive pit. It is adherent to the hypoblast beneath it. Shortly its floor disappears, and its sidewalls open out laterally. Thus it gives rise to the central region of the roof of the archenteron. Its longitudinal axis is the notochord.

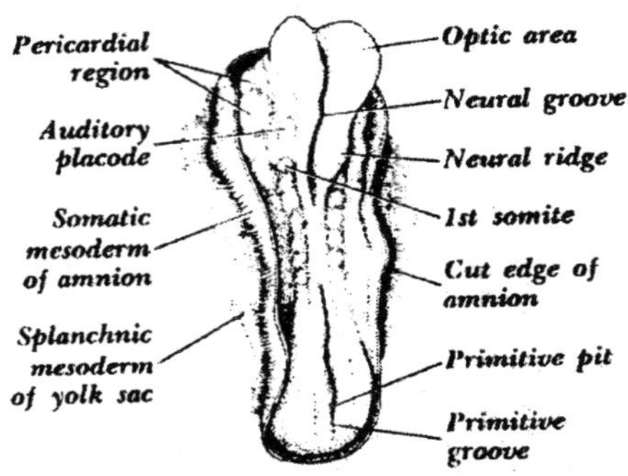

Figure 5.7 : A 7-somite human embryo (2.2 mm, approximately 22 days) seen in dorsal view.

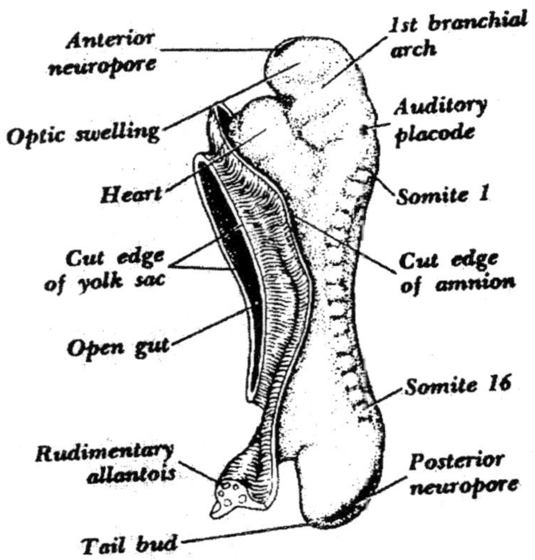

Figure 5.8 : A 16-somite human embryo (2.5 mm, approximately 26 days) seen from the left side. The amnion and the yolk sac have been cut away. At this age the heart has begun to beat. Approximately × 24.

The neural plate forms, and the neural folds rise up and become the neural tube in the mammal, just as they do in the chick. The neural tube pushes forward in the head fold and grows backward in the tail bud, in like fashion. The endoderm develops a foregut, midgut, and hindgut as in the chick. The mesoderm differentiates in the same manner.

The most obvious differences between the mammal embryo and the chick embryo have to do with the relative proportions of some of the organs. The anterior part of the chick embryo is precocious as compared with its posterior portion. The eyes and the brain, especially the midbrain, are exaggerated. In mammals, the development of the heart and liver are surprisingly accelerated. The heart of a 2.5-mm human embryo is bigger than the brain and has already begun to beat. In the chapters which follow, other contrasts between the chick and the mammal will be noted.

TWINNING IN THE MAMMAL

Twinning in the mammalian embryo is interesting and important. Human mothers not infrequently give birth to more than one offspring at the same time. Often these are "fraternal twins" (heterozygous twins). They are actually not true twins at all but "litter mates." They are related to each other as brothers and sisters born on separate occasions. Each develops from a separate egg fertilized by a separate sperm. Although they may be situated close together in the uterus, their membranes are separate.

True, or "identical" twins, also termed homozygous twins, are the products of a single egg fertilized by a single sperm. They have the same chorion and placenta, but their amnions and umbilical cords are usually separate. Since they come from the same fertilized egg, they have an identical heredity.

How do twins of this sort arise? They are probably not the result of a separation of the first two blastomeres, or of the subdivision of the entire blastocyst into two parts. Instead, two embryos arise from the same blastocyst. Either (1) two distinct embryonic discs develop independently on the same blastocyst; or (2) one embryonic disc subdivides and becomes two; or (3) one embryonic disc gives rise to two embryos. In cases (1) and (2) the embryos will have

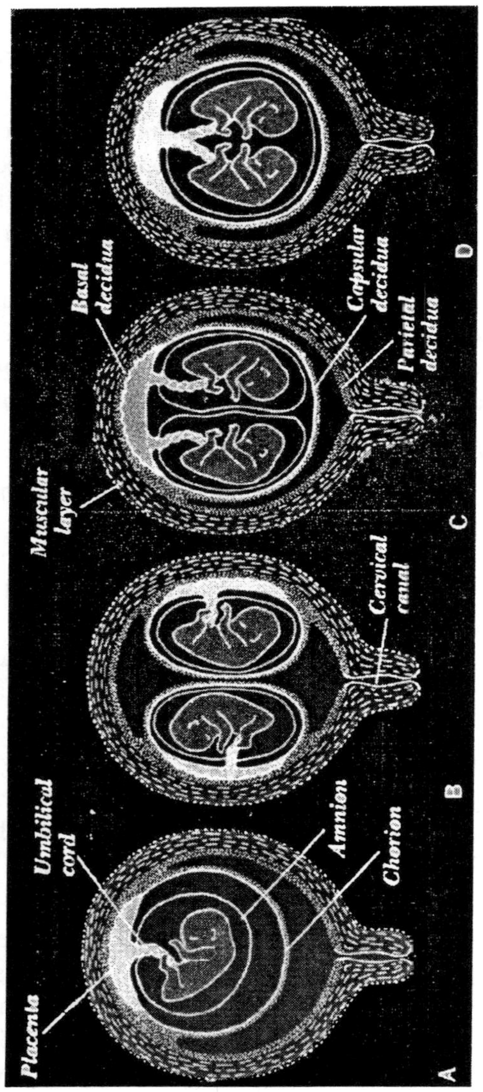

Figure 5.9 : Twinning in man. A. Human uterus with a single embryo. B. Fraternal "twins" (from two eggs) with separate chorions and placentas. C. Identical twins (from one egg) with one chorion and one placenta but with separate amnions and yolk sacs (the latter are not shown). This results when two inner cell masses arise from one blastocyst. D. Identical twins with one set of fetal membranes. This presumably is the result of two embryonic axes forming from one inner cell mass.

separate amnions and umbilical cords. In case (3) it is likely that one amnion will enclose both embryos. Separate identical twins are thought to be usually the result of (1) or (2). Conjoined twins, such as Siamese twins, and the various double-headed monsters, are probably the consequence of (3), that is, the incomplete subdivision of one embryonic disc.

The armadillo is peculiar among mammals in that it regularly gives birth to multiple embryos from one bnlastocyst. It is instructive to think of twinning in vertebrates as a form of asexual reproduction.

6

The Mammalian Embryo

This chapter concerns mammalian parents and their role in the production of offspring. It deals mainly with the adjustments of the mother's body to the embryo and the embryo to the mother.

THE REPRODUCTIVE HORMONES

The story begins when, for some reason, the anterior lobes of the pituitary body of the parents begin to secrete and release increased amounts of gonad-stimulating hormones into the bloodstreams. The time is at puberty, when the young animals commence to become sexually mature. In many mammals the process repeats itself annually, usually in the spring. In any case, it is the prelude to sex behavior. In the case of the female, if reproduction does not take place, the cycle usually begins again at regular intervals.

What causes the anterior lobe of the pituitary to become active? A full answer cannot as yet be given. In some mammals and in birds the increased daylight of spring stimulates the retinas and through them the hypothalamus. This is the part of the forebrain which is close to the pituitary body. Cells of the hypothalamus in turn synthesize a neurohumor which is absorbed into the bloodstream and carried directly to the anterior lobe. The journey is a short one. It involves a set of hypophyseal portal veins which are only about 12 mm long (in man).

When the neurohumor reaches the anterior lobe, it stimulates the lobe to secrete hormones of its own into the blood. There are several of these, each with its own "target organ." The growth hormone, for example, acts on supportive tissues throughout the body, activating them to more extensive growth. The thyrotropic hormone acts on the thyroid gland, causing it to synthesize and release an increased amount of thyroxin into circulation. The thyroxin then stimulates the metabolism of the body. The adrenocorticotropic hormone, ACTH, acts on the cortices of the adrenal bodies, with the result that they secrete hormones which resist stress.

Our present concern is with the pituitary hormones which affect the reproductive functions, namely, the gonadotropic hormones. If the pituitary body of an immature animal is removed, the gonads fail to mature. If the operation is performed after maturity has been attained, the gonads retrogress, and ripe germ cells and sex hormones are not produced. The entire body reverts to a less sexually differentiated state. If, on the contrary, adult pituitary tissue is implanted into a juvenile animal, or extracts of the anterior lobe are injected beneath the skin, or the pituitary develops a tumor, the gonads promptly enlarge and produce germ cells and sex hormones.

The hormones of the anterior lobe of the pituitary which influence reproduction are three in number. Their names and their effects in the female are as follows:

1. The *follicle-stimulating hormone, FSH,* brings about the growth of one or more ovarian follicles and the maturing within each follicle of a ripe ovum. It also stimulates the production by ovarian tissue (interstitial tissue) of the female sex hormone *estradiol.* (The general term for a female sex hormone is *estrogen.)*

2. The *luteinizing hormone, LH,* controls ovulation and the subsequent development of a corpus luteum (see later). Usually it acts in conjunction with FSH. It also stimulates the production and release by the corpus luteum of the hormone of pregnancy, *progesterone.*

3. *Prolactin* is required for the maintenance of the above

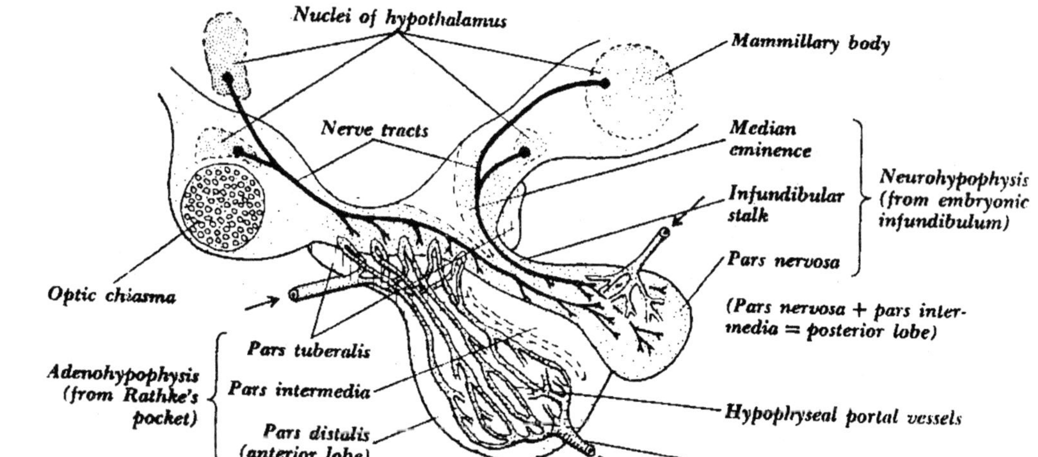

Figure 6.1 : Diagram of an adult pituitary body (median section) illustrating its nerve and blood supply. The adenohypophysis (derived from Rathke's pocket) is shown striated vertically in the case of pars tuberalis, obliquely in pars intermedia, horizontally in pars distalis or the anterior lobe. The neurohypophysis (parts derived from the embryonic infundibulum) is shown stippled. The pituitary is controlled from nuclei (nerve centers) in the hypothalamus. The control of the anterior lobe is indirect: Nerve tracts first carry neurohumors (or their precursors) to the median eminence and pars tuberalis; then hypophyseal portal vessels transport the neurohumors to the pars distalis (anterior lobe) which responds by secreting hormones. The posterior lobe stores and releases neurohumors (hormones) brought to it by nerve tracts from the hypothalamus.

secretions during pregnancy. It also stimulates the growth of the mammary glands and activates parental behavior.

The same three hormones are active in the male, and with comparable effects. In other words, the gonadotropic hormones are not specific with regard to sex.

1. FSH increases the growth of the testes and stimulates the production of sperm cells within the seminiferous tubules.

2. LH causes the interstitial cells of the testes (the cells which lie between the seminiferous tubules) to increase in number and to produce the male sex hormone, *testosterone*.

3. Prolactin affects the male, as it does the female, by calling forth parental behavior. A rooster under its influence will accept and nurture baby chicks.

Biochemists have determined that these several pituitary hormones are, in all cases, proteins or mucoproteins. The hormones of the gonads (and of the adrenal cortex), on the other hand, are steroids. Hormones of a similar kind are present in all vertebrates, although their effects vary somewhat.

We come now to an important principle concerned with the regulation of body processesthe principle of negative feedback. The production of a tropic hormone is antagonized by the hormone whose production it stimulates. Thus thyrotropin is repressed by thyroxin. ACTH is opposed by the adrenocortical hormones. In the same way, the gonadotropic hormones are reduced when the sex hormones of the gonads increase in the circulation. The maximum occurrence of FSH, LH, and prolactin in the blood is found in castrated animals, for in them there are no sex hormones to inhibit the production by the pituitary.

Negative feedback mechanisms result in rhythmic activity, especially if there is a lag between the stimulus and the feedback.

THE OVARIAN CYCLE

The ovarian cycle is an example of a rhythm produced by negative feedback. We begin the story at the time when the reproductive organs of the female are at a low level of activity. The ovary, uterus, vagina, and mammary glands are in as nearly a nonreproductive state as they ever are. No follicle in the ovary has begun to grow. The endometrial lining of the uterus is thin, its blood vessels are reduced, and its secretory activity is at a low ebb. It is then that the sex hormones in the blood are minimal in amount. In primates, including the human female, this state is approximated following the menses. In these, therefore, the cycle is called the *menstrual cycle.* For mammals in general, it might be called the female cycle. (This is not the same as the estrus cycle.)

The events of the cycle, as they take place in the ovary. The figure is schematic, for it shows the state of the ovary stage by stage in a clockwise fashion, beginning at the upper left. Actually, there is no such orderly arrangement of stages in an ovary.

The ovarian cycle divides naturally into two phases: a preovulation phase and a postovulation phase. Each phase is further divisible into stages.

Phase I. The Preovulation or Follicular Phase

STAGE A. Cords of cells bud inward from the germinal epithelium of the ovary into the subjacent connective tissue stroma and there break into "egg nests." Each nest consists of a central oogonium surrounded by a single layer of follicle cells, the *stratum granulosum.* The entire structure is *a primary ovarian follicle.* This process takes place about the time of birth. It is as far as the story goes before puberty.

STAGE B. As puberty approaches, and again at the beginning of each female reproductive cycle, the pituitary body, free of negative feedback, increases its production of gonadotropic hormones, notably, FSH. The FSH stimulates a few-only a few-oogonia to continue their growth as oocytes. At the same time, the surrounding follicle cells multiply until the stratum granulosum has become several cells thick. The majority of the follicles do not complete the process. Instead, they atrophy. In the human female usually only one follicle matures each month.

STAGE C. Under the continuing action of FSH, a cavity appears in the midst of the follicle cells, partially separating an outer

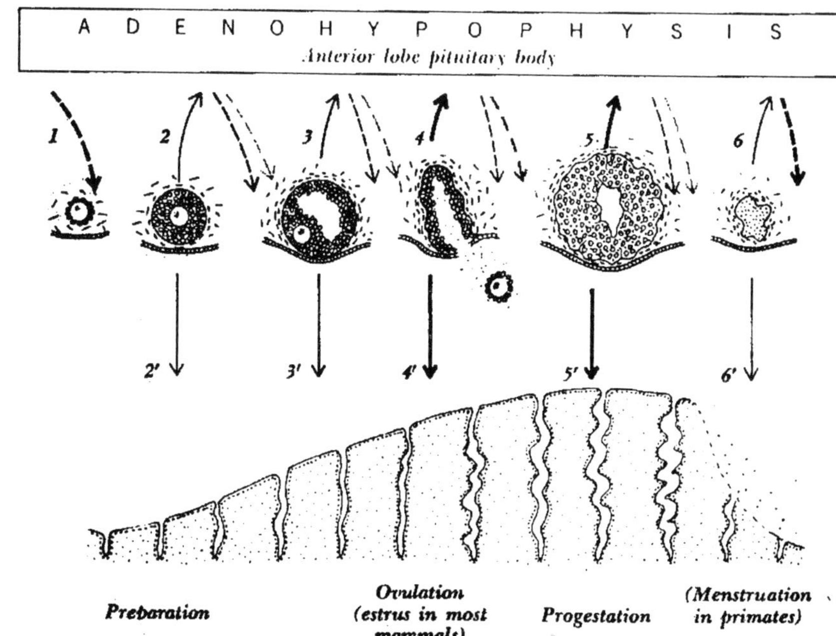

Figure 6.2 : Schema which illustrates the regulation of the ovarian and uterine cycles (human) by hormones of the pituitary body and ovary. 1. The cycle begins with the secretion of FSH (broken black arrow) by the pituitary. This stimulates the growth of an ovarian follicle. 2. The growing follicle secretes estradiol (solid black arrow), which, feeding back on the pituitary body, stimulates the production of LH (broken gray arrow). 2'. The estradiol also stimulates the growth of blood vessels and glands in the uterus. 3. Under the influence of FSH and LH the ovarian follicle matures and produces increasing amounts of estradiol. 4. About the middle of the cycle ovulation occurs, and the follicle cells which remain in the ovary begin to transform into a corpus luteum. The latter secretes the hormone progesterone (solid white arrow). This causes the pituitary to release a third hormone, prolactin (broken white arrow), which prepares the uterus for a possible implantation. 5 and 6. This continues for a time, but if implantation does not take place, the corpus luteum regresses and the uterine mucosa is cast off (menstruation), thus ending the cycle. The breadth of the arrows suggests the relative amounts of hormones released.

layer of cells from an inner mass, the *cumulus,* which lies attached at one side. In the center of the mass is the oocyte. At the same time a sheath of connective tissue cells, the *theca,* forms around the growing follicle and serves to protect the follicle. Its cells also are the principal source of the female sex hormone, estradiol. The latter accumulates in the follicular fluid of the cavity and feeds back to the pituitary gland, decreasing its production of FSH. At the same time, it increases the discharge of LH by the pituitary.

STAGE D. Under the combined influences of FSH and LH, *ovulation* takes place. This is accomplished as follows: The follicle expands until it bulges at the surface of the ovary. Its outer wall becomes thin and finally bursts. The oocyte, surrounded *by a corona radiata* of follicle cells, is swept into the body cavity. In the human female, this ovulation takes place at about the 14th day of the 28-day (menstrual) cycle which starts with the beginning of the menses.

In most mammals, the ripening of an egg is accompanied by a period of heat or *estrus*. Hence the entire cycle from one estrus to the next is termed an *estrus cycle.* Note that this cycle is not the same as the primate menstrual cycle. In some mammals, notably, the cat and rabbit, ovulation takes place only during copulation. Apparently, in these animals an increase of nervous tension is necessary to cause the pituitary body to discharge sufficient LH to bring about ovulation.

Phase II. The Postovulation or Luteal Phase

STAGE E. The increased production of LH and associated FSH by the pituitary body induces the follicle cells which have remained behind in the ovary to multiply, enlarge, and become a temporary gland of internal secretion, known as a *corpus luteum* (a name derived from the Latin, *lutum, a* reddish-yellow dye.) The corpus luteum secretes the hormone *progesterone.*

STAGE F. What follows next depends upon whether or not the egg becomes fertilized and the embryo becomes implanted in the wall of the uterus. If implantation *does* occur, the corpus luteum continues to grow and produce progesterone. As long as the latter hormone is produced, the embryo is retained, no new follicles mature, and ovulation does not again take place. We shall return to this matter under the heading of pregnancy. But if, on the other hand, fertilization and implantation *do not* take place, the growth and activity of the corpus luteum soon reaches a peak, and the secretion of progesterone falls off. The corpus luteum then shrinks to a small mass of scar tissue known as a *corpus albicans*.

These events repeat themselves throughout the reproductive life of the female mammal. In some species they take place without intermission, in others only during the breeding season. The ovary of a young human female at first contains hundreds of thousands of primary follicles. But with each cycle, the number diminishes until at the end of the reproductive period of life very few follicles remain. The rest have atrophied.

UTERINE CYCLE

The reproductive hormones of the ovary control the uterine cycle and the cycles of the other reproductive organs. There are two phases corresponding to the preovulation and postovulation phases of the ovary.

Phase I. During the preovulation phase the uterus responds to the estradiol produced by the ovary by preparing for a possible pregnancy. This phase can be artificially induced by treatment with estrogenic hormones even when the ovaries have been removed or have ceased to function. The endometrium, i.e., the lining of the uterine cavity, thickens as a result of cell proliferation. The blood vessels of the uterine wall multiply, and the glands elongate and become columnar.

This *preparatory phase* continues as long as estradiol is supplied, whether naturally or artificially. If continued long, the uterine tissues hypertrophy (become overdeveloped). It ceases, and the uterus regresses within 6 to 10 days if the supply of estradiol or other estrogen is withdrawn.

Phase II. The postovulation phase of the uterus takes place while progesterone is present in the bloodstream. The two hormones, progesterone and estradiol, acting together, bring about a state which may be thought of as the beginning of a pregnancy, even if, in the absence of an implanted embryo, the pregnancy does not continue. The endometrium grows thicker; the blood vessels enlarge and become distended with blood; and the uterine glands fold back on themselves, become contorted, and secrete a "uterine milk" which is able to nourish an early embryo for several days while it is free within the uterus. For this reason, the postovulation phase is sometimes called the *secretoryphase* of the uterus. In the opossum, implantation of embryos does not occur; the embryos depend upon uterine secretions for their growth.

It is of interest that in some mammals the hormones progesterone and estradiol may so sensitize the endometrium of the uterus that it will respond to a small foreign object by engulfing it.

These changes in the uterus are progressive rather than sudden. In primates they reach a climax at about 10 days following ovulation. If implantation does not take place, the corpus luteum regresses, the level of progesterone in the blood falls, and within two or three days the uterus suddenly returns to its resting condition. This regression is accompanied by menstruation, namely, the sloughing away of the endometrial lining. There is some loss of blood and glandular secretions. The menses may be held in abeyance by the continued use of progesterone. Estradiol or other estrogen alone will not prevent it. These facts have been demonstrated experimentally in animals from which the ovary has been removed.

PREGNANCY

Pregnancy begins when an egg is fertilized and the embryo implants itself in the wall of the uterus. It ends at childbirth. Its duration is spoken of as the "period of gestation." A uterus which is carrying young is termed a "gravid uterus."

In nature, the cycle which includes pregnancy is the usual one. The estral and menstrual cycles are adaptations to a failure of fertilization and implantation.

The timing of the events of the ovarian and uterine cycles is well coordinated. It takes a few days following ovulation for the ovum to descend the oviduct and reach the uterus. During this time the ovum undergoes cleavage. It takes a few more days for the early embryo to reach the blastocyst stage and for its outer protective covering, the zona radiata, to dissolve away. By the time this is accomplished, the endometrium is ready to receive the embryo.

There are a few mammals, such as the armadillo and marten, in which implantation normally is delayed for weeks or even months. In this case, the blastocyst remains inactive in the uterus, waiting for the endometrium to become ready to receive it. One is reminded of the delay, often of several days, which takes place in the bird's egg between the time of laying and the beginning of incubation.

The implantation of the blastocyst in the wall of the uterus has a profound effect on the mother's body. The mechanism of this reaction seems to be that the trophoblast (outer ectoderm of the blastocyst) secretes a hormone which acts on the mother's corpus luteum or on her pituitary body. At any rate, the corpus luteum grows until it is several times the size of the nonpregnant corpus luteum. It secretes increasing amounts of progesterone. The production of estradiol also increases. If the corpus luteum is surgically removed before implantation occurs, implantation does not take place. If the operation is delayed until after implantation has already occurred, the blood vessels of the wall of the uterus regress, the uterine muscles become irritable, and the embryos are rejected.

As the embryo grows, the uterus also grows, as much as twentyfold. Its smooth-muscle fibers increase in number and in size. It is remarkable that this stretching of the muscles of the wall of the uterus does not cause the fibers to contract, for smooth muscle is notoriously subject to stimulation by stretching. The progesterone keeps them quiescent. If progesterone is withdrawn, as by removing the corpus luteum, the muscle fibers become irritable, and the embryo is aborted. This is especially the case during the early stages of pregnancy.

In the late stages of pregnancy in man and monkeys, the removal of the corpus luteum does not result in abortion. Why is this so? Lacking the progesterone of the corpus luteum, what keeps the uterine muscle quiet? The answer seems to be that the placenta itself becomes an endocrine gland which secretes progesterone, and not only progesterone, but an estrogen and a pituitary-like hormone as well. The latter presumably aids in maintaining the production of hormones by the mother's ovary.

THE PLACENTA

A placenta is an organ of attachment between embryonic tissue and maternal tissue. Across it, nutrients and gases are exchanged between the mother and the embryo. In welldeveloped placentas, wastes are transferred also.

Placentas of a sort are present in viviparous fish and in certain reptiles. As a rule, in these cases the yolk sac comes into close relation with the maternal bloodstream, and the vitelline circulation carries the materials to and from the embryo. Yolk-sac placentas are developed in most marsupial mammals and are present also at an early stage in some placental mammals. In occasional reptiles, in a few marsupials, and-most importantly-in all placental mammals, the allantois provides the definitive placenta. It is therefore the umbilical circulation of the allantois which transports substances between the mother and the fetus.

In most instances the area of contact between the mother's uterus and the chorion which encloses the embryo and its membranes is made extensive and intimate by the development of *villi.* These are rootlike processes which grow out from the chorion into the adjacent maternal tissue. Indeed, placentas are commonly classified according to the distribution of these villi.

1. In pigs, horses, and various other mammals, the villi cover most of the surface of the chorion. These animals are said to possess *diffuse placentas.*

2. In cattle, sheep, and other ruminants, the villi are gathered into clusters or rosettes known as cotyledons. Such placentas are termed *cotyledonary placentas.*

3. In cats, dogs, and carnivores generally, the placentas have the form of girdles which surround the embryo and come into contact with the inner wall of the uterus. These are *zonary placentas.*

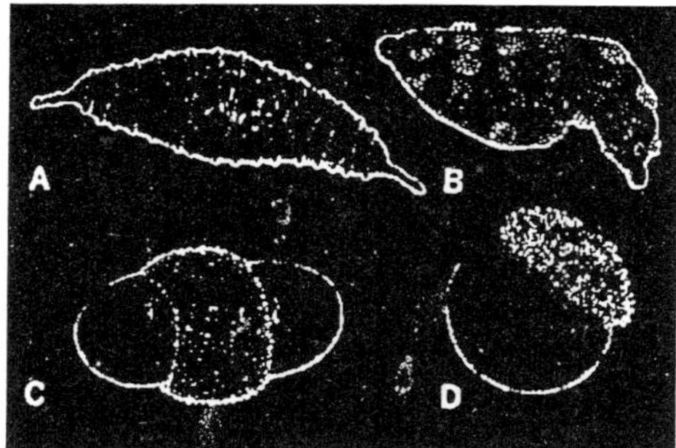

Figure 6.3 : Types of placentas. A. Diffuse placenta (pig). B. Cotyledonary placenta (cow). C. Zonary placenta (cat). D. Discoid placenta (human). At first the entire surface of the human blastocyst is covered with villi.

4. In rodents and primates, including man, the villous area is disc-shaped; hence the term, *discoidal placentas.*

Placentas differ in the intimacy of the contact between the embryonic and maternal tissues. At one extreme, in the pig for instance, the blastocyst elongates and enlarges tremendously. Its entire outer surface comes into contact with the uterine wall and develops villi; but the villi are simple and minute. The relation between the chorionic sac and the endometrium (lining of the uterus) is one of apposition only. In this case, the nutrients and gases which reach the embryo must diffuse across six tissues: (1) the endothelium of the maternal blood vessels, (2) mesenchyme, (3) the epithelial lining of the uterus, (4) the ectoderm of the chorion, (5) fetal mesenchyme, and finally (6) the endothelium of the umbilical capillaries. When at birth *a nondeciduate placenta* of this sort is cast off, there is no loss of maternal tissue or blood.

At the other extreme are the placentas of rodents and primates in which the blastocyst erodes deeply into the endometrium and becomes buried in the uterine wall. The chorion forms complex villi which invade the uterine tissue and rupture maternal blood vessels. When such a placenta is cast off at the time of birth, there is loss, not only of embryonic membranes, but of encapsuling maternal tissue as well. Such a placenta is called *a deciduate placenta.*

The maternal tissues which are expelled at birth in the case of

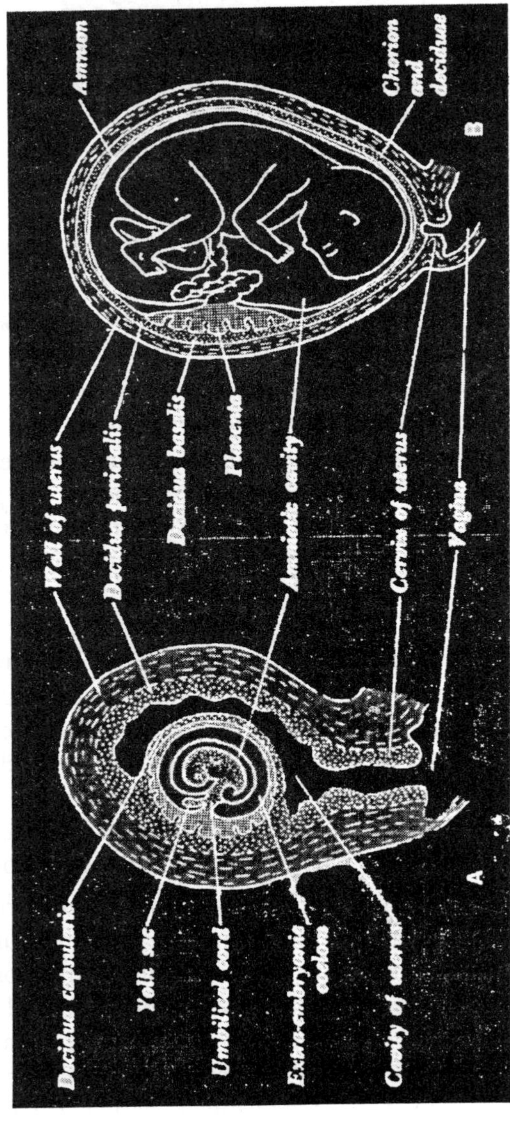

Figure 6.4 : The gravid human uterus. A. At five weeks (8-mm embryo). B. Advanced fetus.

deciduate placentas are called deciduae. There are three regions: (1) The part which lies between the chorionic vesicle and the muscles of the uterus wall is the *decidua basalis.* It alone shares in the development of the placenta. (2) The part which surrounds the chorionic sac and separates it from the cavity of the uterus is the *decidua capsularis.* It is the least developed. (3) The part which forms the inner lining of the rest of the uterus is the *decidua parietalis.*

As the human fetus grows, the cavity of the uterus is squeezed out of existence. The parietal and capsular deciduae come together and fuse, and the latter becomes thin and practically ceases to exist. The amnion becomes closely applied to the chorion. Thus, there is finally only one cavity in the gravid uterus, namely, the amniotic cavity which surrounds the fetus.

The development of the human placenta. Note that the outer ectodermal layer of the blastocyst, namely, the trophoblast, actively destroys maternal tissue as it eats its way into the wall of the uterus. Its outer cells lose their cellular character and become syncytial. Its inner cells, however, remain as a cellular epithelium. The syncytial layer gives rise to complex and branching villi. Within the villi irregular spaces (lacunae) develop and then unite together to form a labyrinth of channels. Through these channels, blood of the ruptured maternal blood vessels slowly flows.

The lacunae split the human placenta into two plates: an outer or *basal plate,* and an inner or *chorionic plate.* Cores of epithelium and mesoderm of the chorionic plate invade the syncytial villi and bring umbilical capillaries into close relation with the maternal blood which flows among the villi between the plates. Late in pregnancy even the syncytial tissue largely disappears, so that little more than the endothelial walls of the umbilical vessels remain to separate the maternal and fetal bloodstreams. This, in brief, is the structure of the human placenta. Details must be sought in more extensive works. The physiology of the placenta is also a matter of great and present interest. It cannot be considered here.

PARTURITION

Parturition is the separation of the fetus and its membranes from

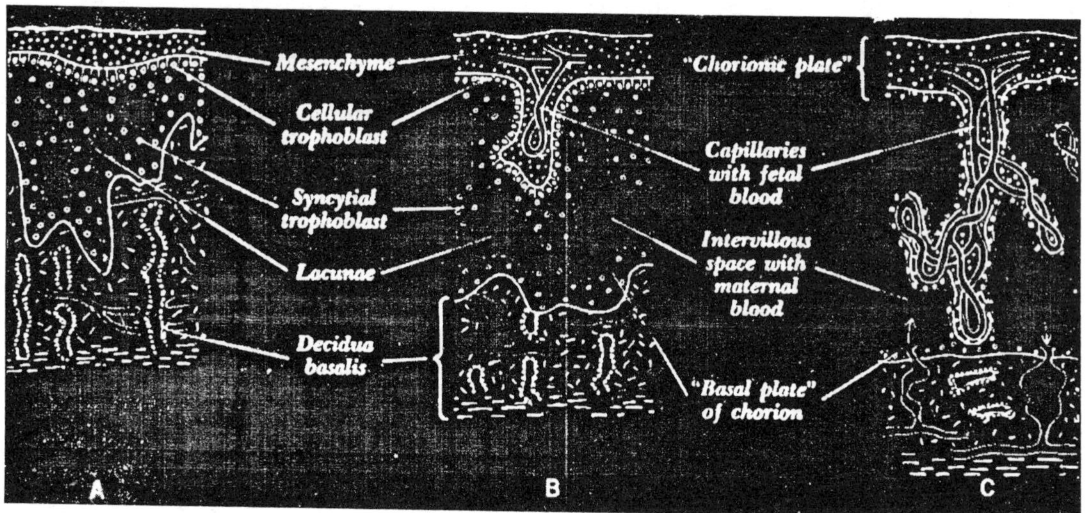

Figure 6.5 : Development of the human placenta (highly diagrammatic). A. The trophoblast (ectoderm of the chorion) has an inner cellular layer (cytotrophoblast) and an outer syncytial layer (syntrophoblast). The latter produces villi which actively erode the endometrial lining of the uterus. B. The villi acquire mesodermal cores which are supplied with capillaries carrying fetal blood. Lacunae in the syntrophoblast unite and become intervillous spaces filled with maternal blood. The spaces divide the chorion into a "chorionic plate" and an outer "basal plate" next to the uterine tissue. C. The villi become very complex (bushy). The trophoblast covering the villi finally degenerates until very little remains separating the maternal and fetal bloodstreams except the endothelium or the villous capillaries.

the mother's body at the time of birth. Those aspects of birth which have to do with the newborn, such as the first breath and the reorganization of the circulation, will be described in connection with the respective organ systems. We are now concerned with the mother.

A hormone-like substance termed *relaxin* appears in the bloodstream as the period of gestation draws to a close. This substance seems to be a product of the uterus or of the placenta, and it serves to prepare the mother's body for the work of childbirth. Hisaw discovered it first in guinea pigs as being a substance which relaxes the pubic symphysis. It has since been reported in our domesticated animals and in man. In man it renders the cervix of the uterus more readily dilatable.

As gestation draws to a close, rhythmic contractions of the smooth muscle of the uterus begin. Soon they are accompanied by reflex and voluntary activity of skeletal muscle. The result is that internal pressure forces the fetus, normally head first, into the cervix of the uterus. The fetal membranes (combined deciduae, chorion, and amnion) then bulge through the cervix and burst. Amniotic fluid is discharged. The child then enters the vagina and is born. Some 15 minutes later, the uterus again goes into rhythmic contraction, and an "afterbirth" consisting of the fetal membranes and deciduae is expelled. Surprisingly enough, there is usually only a minor amount of bleeding. Within the next few days, the endometrium is restored and the uterus returns to nearly its original dimensions.

The mechanisms of birth are complicated and not well understood. Nervous reflexes of the lower cord are involved. Nevertheless, parturition will take place even when the spinal cord is transected. Higher centers are also involved. *Oxytocin, a* hormone of the posterior lobe of the pituitary gland, when given hypodermically, augments the contraction of uterine muscle at childbirth; yet delivery will take place even in animals from which the pituitary body has been removed. It is not entirely certain, therefore, that oxytocin is normally involved in the birth of the child.

What activates the nerve centers? What decrees that the time for birth has come? The answer is hormonal. Supposedly there is a decrease in the amount of progesterone in the bloodstream. We have seen that during late pregnancy progesterone is produced by the placenta. It maintains the uterus in a quiescent state. If the fetuses and placentas of a pregnant rat are removed and replaced by paraffin balls, the balls are expelled within two days. But if some of the placentas are left in place, parturition does not take place until the normal time. If the fetuses of a mouse or monkey are destroyed within the uterus without damaging the placentas, the placentas are retained and continue to produce placental hormones until full term. These experiments indicate that the placenta has a definite span of life.

Genetic studies on cattle indicate that the hormones of the fetus (including its placenta) rather than those of the mother set the time for delivery. A cow which regularly gives birth to normal calves at the proper time-280 to 290 days-will carry a certain type of abnormal fetus for months beyond the expected period of gestation. The abnormality in question results when the fetus inherits a double dose of a recessive gene. It shows itself in defects of the pituitary body and deficiencies in the pituitarylike hormones 'of the fetus. There is still much to be learned, however, about the factors which bring about childbirth.

LACTATION

The mammary glands of juveniles and males are mere rudiments. During adolescence and under the influence of estrogens, the glands of young females grow by proliferation of the ducts. There is some waxing and waning with each recurring female cycle.

During pregnancy, the mammary glands increase greatly and become functional. The first half of each pregnancy is marked by the multiplication of the ducts and the development of acini, namely, the terminal secretory sacs. This takes place in response to various pituitary and ovarian hormones but especially the progesterone secreted by the corpus luteum. Yet the progesterone of the placenta is even more important, for if the placentas of a pregnant rat are removed, recession of the mammary glands soon follows. Recession does not occur if the corpora lutea or even the fetuses are taken away and the placentas are left in place.

During the second half of pregnancy, the acini become distended with secretion. The agent operating here is prolactin, produced by the pituitary. It is of interest that under the influence of prolactin the crop glands of a pigeon produce a "milk." Prolactin also stimulates parental behavior and nest building in birds and mammals.

By treatment with a proper sequence of hormones, male mammary glands can be made to secrete milk and the male animal caused to accept and suckle young.

What inhibits the flow of milk from the mammary glands before birth, and what brings forth the flow of milk at the time of birth? Probably gonadotropic hormones are responsible, especially prolactin. But the nervous reflexes associated with suckling are also important.

7

Development of Rabbit

Mammals evolved from reptiles. Some mammals still lay their large eggs like most modern reptiles and birds. These animals are called the *viparous animals* and such condition of reproduction is called *oviparity*. Most mammals give birth to young ones directly and in them the development is internal and takes place inside the body. Such animals are the *viviparous animals* and the stage is called *viviparity*. The viviparous animals are those animals in which the eggs are internally fertilized, the fertilized eggs are retained in a specialized part of oviduct called *uterus,* and the embryological development of these eggs take place inside the uterus. The developing embryo of such a viviparous animals often establishes a direct connection with the uterine wall and draws its nutritional, respiratory, excretory and other metabolic requirements from the maternal body. All mammals, except prototherian mammals which are oviparous (e.g., *Ornithorhynchus)* or *ovoviviparous* (a reproductive condition of egg-lying animals, in which the eggs though develop inside the maternal body, but are not supplied with maternal nourishment, e.g., *Echidna),* are viviparous. The viviparity is an adaptation to adverse environmental conditions such as cold, dryness and seasonal changes. It permits the. mother freedom to move about, to feed, and to migrate, a freedom which is not enjoyed by a bird parent during its task of incubation. Viviparity is also found in some elasmobranchs, s teleosts and reptiles. In most of these animals, however, the egg is supplied with abundant yolk, but develops within organs of its mother's body, until hatching.

EVOLUTION OF VIVIPARITY IN MAMMALS

Palaentological evidences indicate that the mammals are derived from ancestors which were closely related to the early reptiles. It is expected, therefore, that they must have had large yolky eggs and their developmental history must have resembled with that of the reptiles. At some stage of their evolution some of the mammals switched from oviparity to vivipary. In such cases embryo received sufficient supply of nourishment from the mother while it was retained in the uterus, and the yolk supply of the egg became unnecessary. As a result, there was a progressive decline in the yolk supply and eventually it disappeared altogether. In this respect, there is a distinct graduation in the three sub-classes of the class Mammalia.

1. Sub-class Prototheria

Mammals of this sub-class are mostly *oviparous* as they lay eggs that develop outside the maternal body, e.g., *Ornithorhynchus.* However, some of them are *ovoviviparous, i.e.,* through developing in the maternal body, they are not supplied with nourishment, e.g., *Echidna.* The eggs, therefore, have a large amount of yolk. They are meroblastic and develop in a reptilian pattern.

2. Sub-class Metatheria (the Masupials)

The quantity of yolk in the egg is insignificant and it becomes unnecessary for the development of the embryo. Nevertheless, it is ejected at the beginning of cleavage. Although the developing embryos receive nourishment from the mother in the uterus, but the adaptations for this procedure are not much advanced and the young are born poorly developed.

3. Sub-class Eutheria (Placental Mammals)

The microlecithal egg is devoid of yolk right from the very beginning. Therefore, the embryos of these mammals develop inside the uterus and draw their nutrition from the maternal body through a 'physiological bridge' the *placenta.*

THE SPERM

It is microscopic. It is formed of a head, a middle piece and a tail. The head is pear-shaped. It has a nucleus and an acrosome. The nucleus and acrosome are surrounded by a plasma membrane. The middle piece is connected with the head by a neck The middle piece contains two centrioles and the mitochondria. The tail is long. Its central core is occupied by an axial filament.

THE EGG

The egg is spherical in shape it is microscopic and is about 0.11 mm. in diameter. The smaller size of the egg is due to the absence of yolk. Since there is no yolk, the egg is said to be *alecithal.* When the egg is released from the ovary it is surrounded by three membranes namely, corona radiata, a middle zona pellucida and an inner plasma membrane. The corona radiata is formed of folicle cells. The zona pellucida corresponds to the vitelline membrane of lower vertebrates. There is a fluidfilled space between zona pellucida and the surface of the egg and it is called *perivitelline space.*

Mammalian eggs are Secondarily Alecithal

The egg of Amphioxus has lesser amount of yolk (alecithal) like that of mammals. The absence of yolk or the presence of lesser amount of yolk is a primitive condition as it is exhibited by the primitive chordate. The amount of yolk increases in the case of the animals which are highly evolved. Thus fishes, amphibians, reptiles and birds are provided with a large amount of yolk in their eggs. The amount of yolk reaches its maximum in birds. Among the mammals, the amount of yolk in prototheria is as high as in birds. But the eutherian mammals (man, rabbit, pig etc.) possess only a negligible amount of yolk (alecithal). Thus the egg of highly advanced mammals resembles the egg of primitive chordate, *Amphioxus.* The alecithal condition of mammalian egg is due to the internal development of the embryo. The mother provides all the necessary materials for the development of the embryo. That is why yolk is not deposited inside the mammalian eggs. Thus the mammalian eggs are not *primarily alecithal* but only *secondarily alecithal.* The Amphioxus-egg is primarily alecithal. The alecithal egg of mammals is derived from the megalecithal egg of reptiles and birds. In the course of evolution, the mammalian eggs became alecithal only because of internal development.

The fact that mammalian eggs are derived from megalecithal eggs is further confirmed by the pattern of development especially gastrulation. The gastrulation of mammalian egg is exactly like that of megalecithal eggs of reptiles and birds and not like that of alecithal eggs of *Amphioxus.* Again like megalecithal eggs, the mammalian eggs develop an *yolk sac.*

FERTILIZATION

Fertilization takes place in fallopian tube. During copulation, the penis is inserted into the vagina; the sperms thus transferred by the male, travel up the oviduct and reach the fallopian tube to fertilize the eggs. Complete sperm enters the ovum; the tail, however, soon degenerates. Soon after the entry of the sperm, the egg completes its maturation. The fertilized egg passes down the oviduct into the uterus where it becomes attached to the uterine wall. A vascular organ, the *placenta is* formed through which developing foetus gets the nourishment from the maternal blood. The period of *question (*= period of carrying the young within the uterus) is normally one month. When fully formed the uterus expels the young ones by muscular contraction at birth. Rabbit is viviparous and the young-born possesses all the organs fully formed. The Female rabbit gives two young ones at a time. At birth the young one is hairless and its eyelids are coherent.

CLEAVAGE

The fertilized eggs or rabbit and other entherian mammals are *microlecithal,* minute in size, i.e., ranging in diameter from 0.08 to 0.15 millimetre. Cleavage of the eggs of rabbit is of *holoblastic* type, but despite a microlecithal condition the blastomeres tend to show size differences from the start. Moreover, the numbers of blastomeres do not increase by a regular doubling sequence but tend to show arithmetical progression. The cleavage process of rabbit culminates in blastula by undergoing following stages:

First Cleavage

The first cleavage is *meridional* or vertical and takes place within 22 to 24 hours after mating or 10 to 12 hours after fertilization. This results in the production of two unequal blastomeres. The smaller one is called *micromeres* and the larger is called *mega* or *macromere.* The blastomeres are held together by the zona pellucida.

Second Cleavage

The second cleavage is also called *meridional* or *vertical,* but at right angles to the first one. The four cell stage is reached about 24 to 32 hours after fertilization. Although total, the cleavage *is irregular* as the synchronization of untosis in the blastomeres is lost very early. The macromere divides first and thus a three cell stage is produced. This is followed by the division of the micromere. Thus a four-cell stage is formed.

Further cleavage divisions are irregular. The *eight-cell stage* is found after 32 to 41 hours of mating. One member of the larger blastomeres of the four-cell stage egg divides, forming a five-cell condition, which is followed by the division of second larger blastomere, producing a six-cell stage. After a short period, one of the smaller cell divides and thus, a total of seven blastomeres is formed. The last cleavage is followed by the division of the other smaller cell, producing a eight-cell stage. The mitotic spindles of each of these cleavages are formed at right angles to one another, thus, showing an independence and asynchrony of cleavage. The rate of mitotic divisions increases and at about 65 to 70 hours after mating, the 16-cell stage is reached.

MORULA

Cleavage results in the production of a solid mass of cells. This mass of cells appears like a bunch of mulberry. Hence this stage of the embryo is known as *morula.* It has no cavity inside. It is formed within 65 to 70 hours after mating: The morula consists of two types of cells, the micromeres and macromeres. The macromeres became flattered and form a covering. It is called *trophoblast* or *trophectoderm.* It is nutritive in function. The interior of morula is occupied by the micromeres.

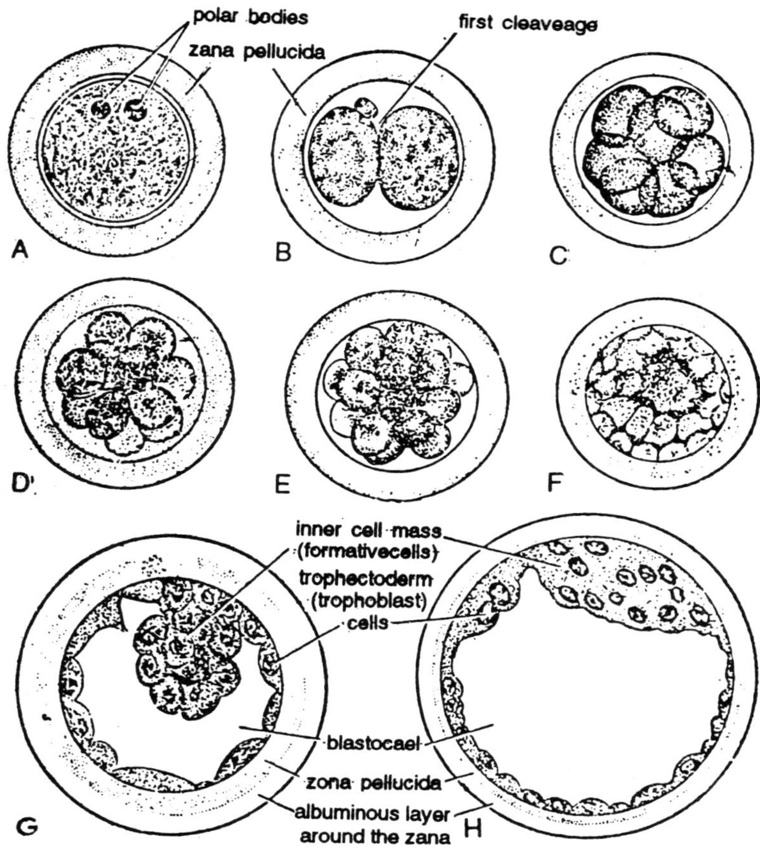

Figure 7.1 : Early cleavages of rabbit.

They are called *inner cell* mass. The inner cell mass alone develops into the body proper of the embryo.

BLASTOCYST

Soon a cavity appears inside the cell mass on one side; this cavity gradually increases. This results in the shifting of the central mass to one side which eventually becomes attached to the outer layer at the animal pole. The central mass is now called the *inner cell mass* while the cavity is falsely called the blastocoel, filled with the fluid imbibed from the secretion of the uterine wall. This stage can be referred to us *blastula* and is called *blastocyst* or *blastosphere* in mammals. By the further divisions of the inner cell mass, the blastocyst, grows to the size of 0.28 mm. diameter with inner cell mass attached at *embryonic knob* at the animal pole. From the embryonic knob, the embryo arises. The general outer layer of cell which encloses the blastocoel and the knob is called the *trophoblast.* It soon establishes relations with uterine wall and helps in the nutrition of the developing embryo. The cell of the trophoblast overlying the embryonic knob, are called the *cells of Rauber.*

IMPLANTATION

The embryo reaches the uterus in the blastocyst stage. Then it becomes attached to the uterine wall. The process by which the blastocyst is attached to the uterine wall is called *implantation.* Implantation is really the adhesion of the trophoblast of the embryo to the uterine wall. The adhesion occurs selectively where there is maternal blood vessels beneath the uterine epithelium. By this attachment the maternal blood diffuses into the embryo.

Types of Implantation

In mammals there are three types of implantation. They are (a) central or superficial implantation(b) eccentric implantation and (c) interstitial implantation.

(a) *Central or Superficial Implantation*

In central implantation the blastocyst remains within the cavity of the uterus;- the trophoblast (outer layer of blastocyst) makes a superficial attachment to the uterine wall (epithelium), e.g., Ungulates, pig, carnivores, monkey etc.

(b) *Eccentric Implantation*

Here the blastocyst remains in between the folds of -the uterine epithelium. Then the fold s grow over so that the blastocyst is no more in the cavity of the uterus, e.g., rat ; squirrel etc.

(c) *Interstitial Implantation*

Here the blastocyst burrows deep into the uterine wall e.g., hedgehog, guinea pig some bats, man etc.

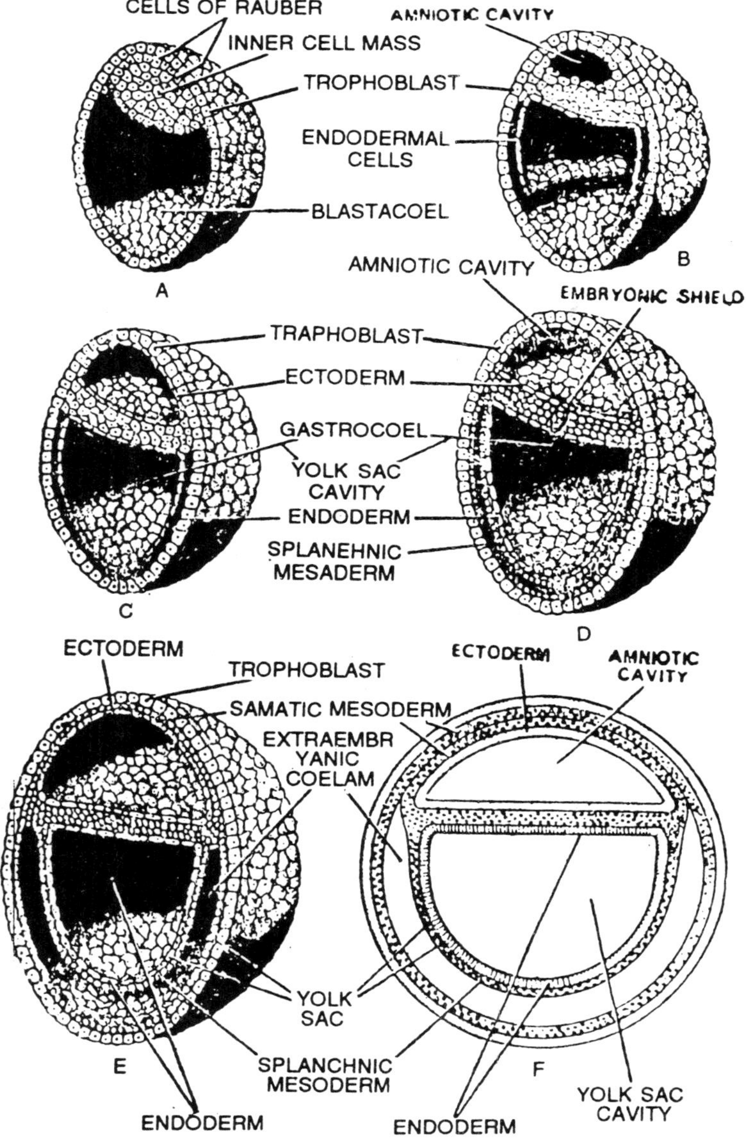

Figure 7.2 : Development of Rabbit, Gastrulation of the mammalian embryo. The inner cell mass forms the amniotic cavity by delamination. The embryonic shield is located between the amniotic cavity and the archenteron.

GASTRULATION

Gastrulation begins either after implantation or earlier. There are two steps as in chick They are

1. The development of endoderm and
2. The development of primitive streak and mesoderm.

1. The Development of Endoderm

The endoderm or hypoblast is formed by delamination. The innermost layer of cells of the inner cell mass (cells facing the blastocoel) splits off from the remaining cells. The delaminated cells multiply rapidly and gradually extend downward along the inner surface of trophoblast. They meet and fuse on the midventral side enclosing a cavity. This cavity is called *yolk sac* (archenteron) because it is homologous to the yolk sac of birds. In birds this sac contains yolk; but in mammals there is no yolk Even then, the yolk-sac develops as an ancestral memory. After the formation of hypoblast, the inner cell mass is called *embryonic knob.*

As the hypoblast is being formed, the cells of Rauber covering the inner cell mass disintegrate and the inner cell mass is exposed to the exterior. At the same time, the embryonic knob proliferates rapidly and the cells arrange themselves in the form of a disc called *embryonic disc* or *germ disc.* The embryonic disc is connected with the trophoblast around its edge. The upper layer of cells of the embryonic disc develops into the embryonic ectoderm and the trophoblast is now called trophectoderm.

Fate Map

The organ-forming areas (fate-map) are arranged in the embryonic disc as in chick.

2. Development of Primitive Streak

The second step in gastrulation is the development of primitive streak and the subsequent formation of mesoderm. After the establishment of the endoderm, the embryonic disc at one side proliferates cells at a rapid rate and hence the cell number increases in this region. This results in a definite increase in thickness of the embryonic in this region. The thickened area is crescentic in shape when it first appears. It is called the primitive streak The side, where the primitive streak appears, represents the caudal end of the embryonic disc. The opposite end represents the cephalic end where the head develops later. Soon the primitive streak elongates cephalo caudally. This ;is followed by the over all lengthening of the embryonic disc. The primitive streak consists of primitive plate, primitive ridges, primitive groove, primitive pit and Hensen's node as in chick.

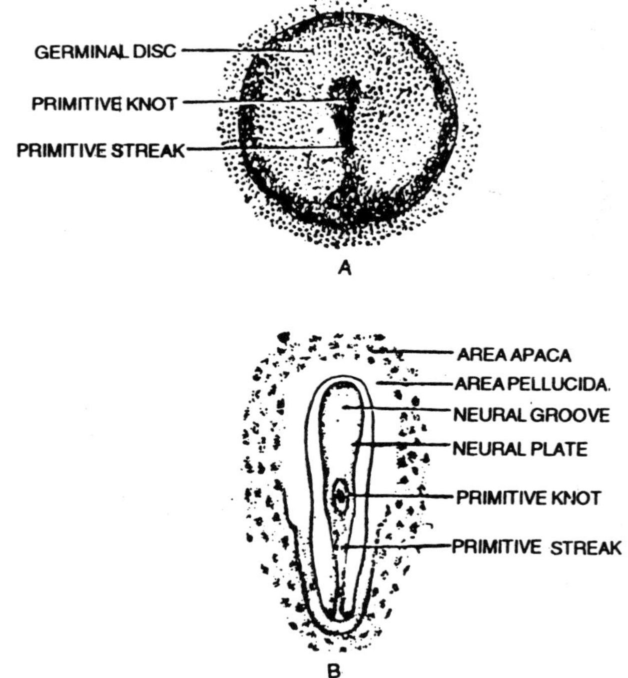

Figure. 7.3 : Development of mammals. A-Appearance of the primitive streak and groove in the blastodisc; B-Surface view of an early embryo.

As the primitive streak is growing, its cells migrate downwards into the blastocoel. These are the mesenchyme cells. Once inside, they migrate laterally as well as forwards between ectoderm and endoderm. Here these cells arrange into two sheets on either side of primitive streak. They form the mesoderm. The mesoderm cells migrate outside the germ disc also. The mesoderm which is found

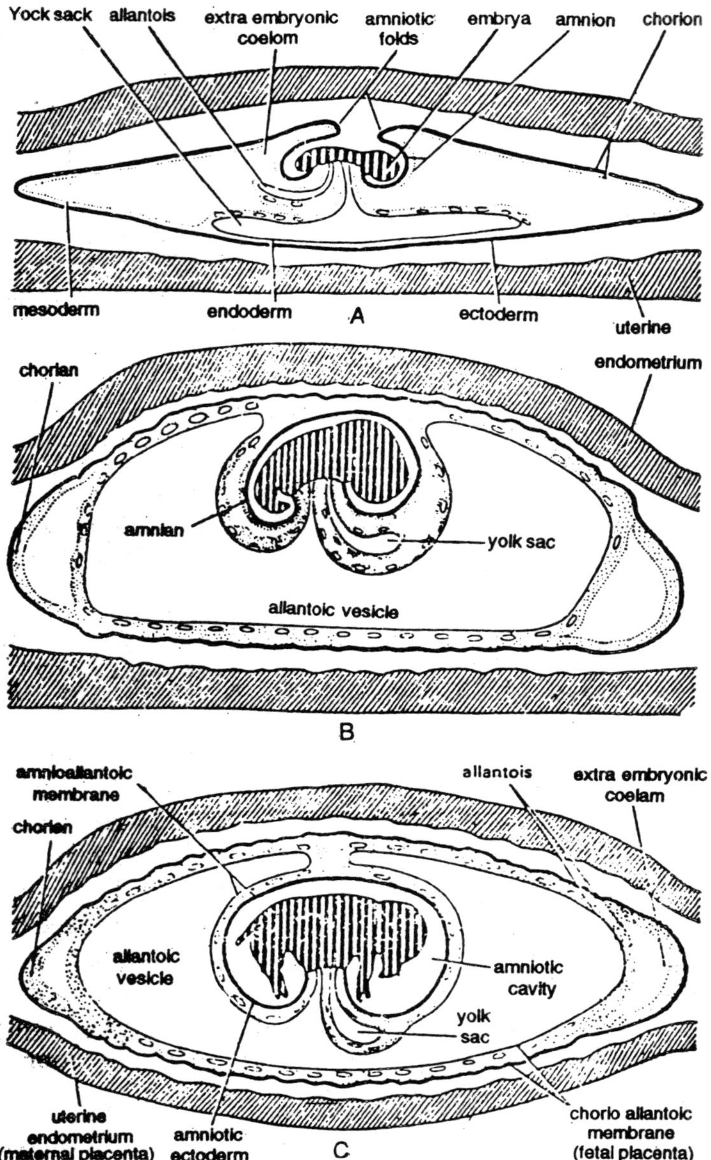

Figure 7.4 : Development of extra-embryonic membranes of pig. A-B—Early stages. C—Fully developed extra-embryonic membranes.

inside the germ disc is called, *embryonic mesoderm* and that lying outside the germ disc is called *extraembryonic mesoderm*. The extraembryonic mesoderm develops into the extra embryonic structures like foetal membranes. The mesoderm undergoes the same type of changes as in chick.

The cells migrating into the blastocoel from the Hensen's node, arrange themselves in the form of a rod in front of Hensen's node. The anterior portion of this rod is called *prechonial plate* and the posterior portion is called *notochord*.

The ventral side of notochord is in close contact with the endoderm. Sometimes, in this region, the endoderm disappears completely and the roof of archenteron is formed of notochord only. The posterior end of notochord makes contact with the Hensen's node. As the primitive streak recedes pact wards, the notochord stretches backwards. When the primitive streak disappears its place below the epiblast is taken by the notochord. The neural tube, noto-chord, mesoderm and coelom-develop as in chick.

FOETAL MEMBRANES OF RABBIT

The mammalian membranes including rabbit develop four *extra embryonic membranes* or *foetal membranes* to cover and the protect the embryo. These are *anmion, chorion, allantois* and *yolk sac*. The structure, development and function of these membranes are more or less similar to those of chick and slight modification.

AMNION AND CHORION

Amnion and chorion develop as upward projecting folds of somatopleure (somatic mesoderm and trophectoderm) called *amniotic folds. Two* types of amniotic folds develop. They are named according to their positions. They are the amniotic head fold, and the amniotic *tail fold*. First of all the amniotic fold appears as a transverse fold behind the *tail*. It is called *amniotic tail fold*. It grows over the embryo dorsally and then grows forwards. Later a transverse fold appears in front of the head. It is called amniotic head fold. It grows upwards and the backwards over the embryo. Finally the head fold and the tail fold meet and fuse together. This fusion takes place near the anterior end because the tail fold develops first and grows forward rapidly. The place of their final fusion is marked by a temporary connection called *seroamniotic connection* or *seroamniotic raphe.* After the union of the folds, the inner and outer layers separate. The inner layer becomes the amnion and the outer layer becomes the *chorion. For* further development refer to the chick.

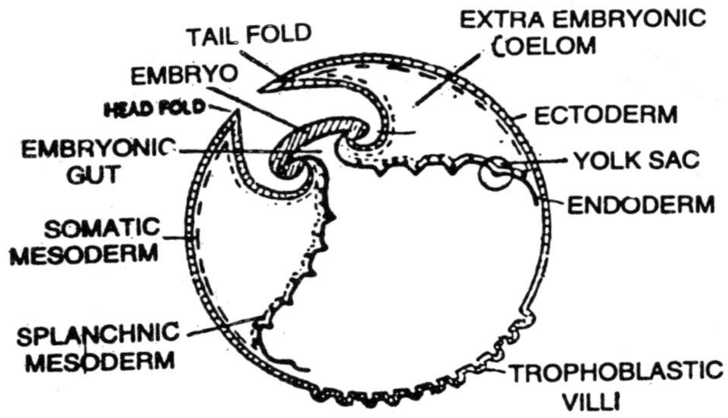

Figure 7.5 : Diagram showing formation of amnion and chorion.

Functions

1. The embryo is freely immersed and bathed in the amniotic fluid. Hence the amniotic fluid functions as the *artificial swimming pool of the embryo.*

2. As the amniotic fluid is more viscous and gelatinous, it functions as an efficient shock absorber. Thus the soft embryo is protected from external as well as internal mechanical pressures.

3. The amnion prevents the adhesion of embryo to the other parts.

4. The amnion provides facilities for the free movement of the embryo.

5. The chorion provides psysiological exchange between the embryo and the mother. It develops finger-like outgrowths called *villi.* The villi dip into the corresponding depressions called *crypts* developed in the uterine wall. This enhances the absorption of nutrition from the maternal side.

YOLK SAC

There is no yolk in mammalian eggs. Hence the yolk sac is not necessary for mammalian embryos. Even then the yolk sac develops, as it developed in the ancestors (reptiles)). Hence the development of yolk sac in mammals is a mere repetition of an ancestral character. So the yolk sac in mammals is empty.

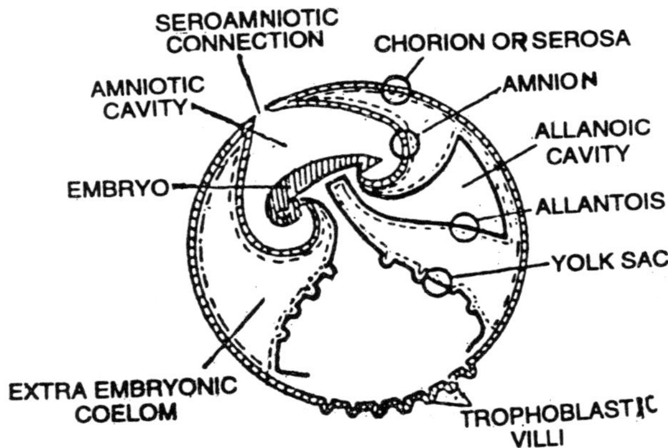

Figure 7.6 : Diagram, showing formation of yolk sac placenta.

The yolk sac initiates its development, when gastrulation begins. In the blastocyst stage, the inner cell-mass liberates some cells into the blastocoel. These cells arrange themselves to form a flatened layer below the inner cell-mass. This layer forms the *endoderm or* the hypoblast. As more and more cells are added to this layer, the endoderm grows downward along the inner surface of the trophoblast. The space enclosed by endoderm is called the yolk sac. In the early stage the yolk sac is lined by the endoderm only. But when the mesoderm develops, the splanchnic mesoderm grows in close association with the yolk-sac endoderm. The mesoderm will not extend to the lower part of the yolk sac. The upper part of the yolk sac is highly vascularised and contains vitelline arteries and veins. The lower part of the yolk sac is not vascularised.

The connection of yolk sac with the embryo becomes narrowed into a tubular stalk called *yolk stalk.* Later thee yolk sac gradually decreases in size and becomes shrivelled.

The yolk sac absorbs uterine milk and the milk is transported by the vitelline circulation to the embryo as nourishment. Thus the yolk sac is nutritive in function. In addition, it assists respiration. Blood cells are also produced by splanchnopleure of the yolk sac.

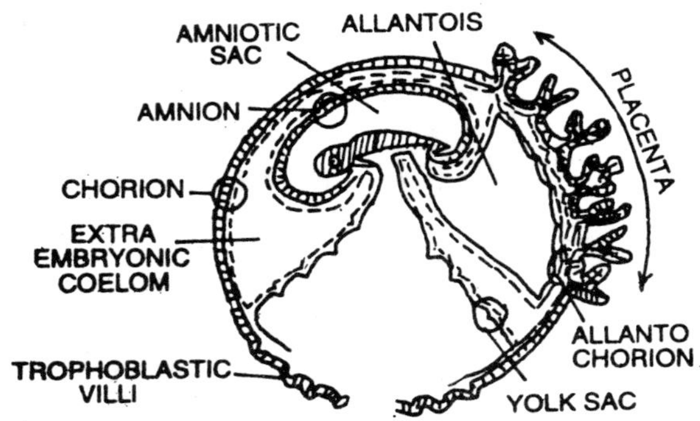

Fig. 7.7 : Diagram, showing formation of placenta.

ALLANTOIS

The allantois of rabbit and other mammals is composed of splanchnopleure (endoderm + splanchnic mesoderm). It arises from the hind gut as a small sac which gradually increases in the extra-embryonic coelom, i.e., space between the amnion and chorion. Soon, it grows over the embryo outside the amnion where it enlarges. Later on, mesodermal component of allantoic wall give rise to the blood vessel system of the allantois and fuses with mesoderm of chorion to form of *allantoic chorion.*

Functions of Allantois

With the. acquisition of viviparity in mammals, the original function of the allantois as a urinary bladder becomes altogether lost. As the maternal organism supplies the embryo with all the necessary nutrition,, so also it takes over the removal of the waste products (CO_2, urea, etc.) of metabolism from the embryo. However, allantois of mammals supplies oxygen and nutrients to the embryo, which are taken from maternal blood circulation.

The Nutrition, Respiration and Excretion of the Developing Embryo and Foetus

When the developing blastocyst reaches the uterus, the lower region of the trophoblast develops villi which penetrate into the uterine wall. The uterine wall at this time becomes more vascular i.e., its blood vessels dilate. At this stage the blastocyst is nourished by the secretion of the uterine wall and by the food absorbed from it by the trophoblast villi. With the formation of foetal membranes, larger finger-like viii arise from the allanto-chorion and the embryo sinks deeper into the decidua formed by the multiplication of the cells of the uterine epithelium. Thus a firm connection develops between the foetus and the mother through *placenta.* The placenta is partly maternal and partly embryonic. The chorionic villi soon become more vascularized to form *placental villi* containing circulation which is continuous with that of embryo via allantoic blood vessels and capillaries connected with the extra-embryonic arteries and veins. The tissues of the uterine wall in contact with these villi break to form lacunae through which maternal blood flows to bathe the villi. This facilities the diffusion of food and oxygen from the maternal blood to that of foetus while the waste products and carbon dioxide produced by foetus diffusein the reverse direction. Thus we can say that the foetus utilizes the digestive system, lungs and kidneys of the mother.

Soon the heart of the foetus becomes four-chambered with the pulmonary circulation still wanting as the lungs are expended. Therefore the blood returning to the right auricle does not pass to the lungs but is largely short-circuited into the left auricle through *foramen ovale.* This aperture lies in the partition that separates the two auricles. From left auricle the blood goes to the left ventricle and then to the general circulation.

At the time of birth, while the foetus is passing, the foetal membranes burst with the result the young burned rabbit possesses no enclosure of foetal membranes. The young rabbit takes the first breath, expands the lungs thereby establishing the pulmonary circulation. In the meantime foremen ovale and ductus arteriosus close and thus a true double circulation is established.

8

Development of Pig

The maturing eggs of the pig are expelled from the ovary during the period of heat, following which they become promptly fertilized.

As soon as the furtilization is completed the development begins. cleavage, blastulation and even the gastrulation occurs in a typical mammalion pattern.

At the completion of germ-layer formation the embryonic disc possesses a typical primitive streak. This is quickly followed by the appearance of neural folds and then mesodermal somites. The stages immediately succeeding correspond to those of three-day chick embryos, but are complicated by flexion and spiral twisting this makes sections difficult for the beginner to interpret. However, in embryos about 6 mm. long the twist of the body has disappeared sufficiently so that its structure may be studied to better advantage. At this time the state of development is generally comparable to that of a four-day chick.

The fetal membranes of the pig stand somewhat intermediate between those of the chick and man. The amnion, chorion and allantois develop very much as in the chick. The yolk sac for a time grows rapidly, but its functions are soon transferred to the allantois which fuses with the chorion; the two enter into a contact-relation with the lining of the uterus and constitute a placenta, which is the organ of fetal respiration, nutrition and excretion. The develo-pment and relations of these extra-embryonic structures are described.

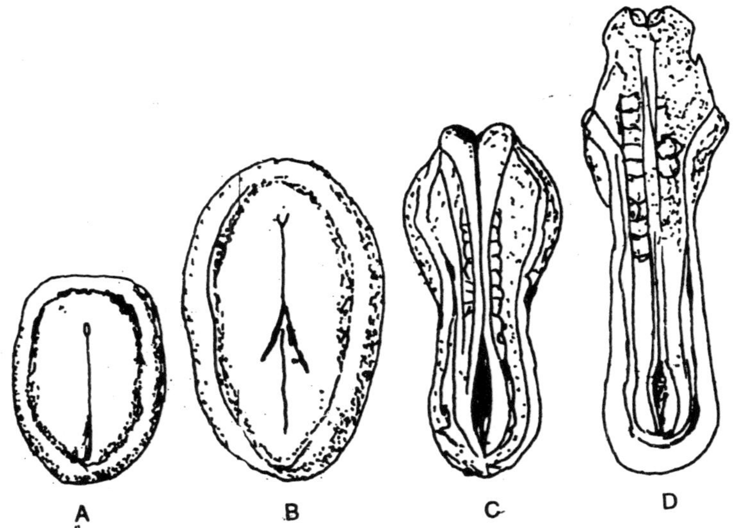

Figure 8.1 : Early pig embryos, in dorsal vie. Blastoderm at twelve days, with primitive streak and knot. B, Blastoderm at thirteen days, with primitive streak and neural groove. C, Embryo of fourteen days, with seven somites. D, Embryo of fifteen days, with 11 somites.

A. Anatomy of A 6 m.m. Pig Embryo

The general structure of 6 mm. pig embryo is illustrated in Figures. This should be compared with the chick embryo of four days and the 5 and 8 mm. human embryo. Some familiarity with the gross anatomy of the 6 mm. pig embryo will make the detailed study of the 10 mm. pig embryo, which follows easier to understand.

B. Anatomy of a 10 m.m. Pig Embryo

This is the most instructive single stage of later development. Nearly all the important organs are represented, and yet the embryo is not so complex as to confuse unduly a beginner. Embryos between 8 and 14 mm. long may be used satisfactorily in conjunction with the descriptions that follow. At this period a human embryo is slightly further advanced than a pig embryo of equal size, but at corresponding stages of development they are fundamentally alike.

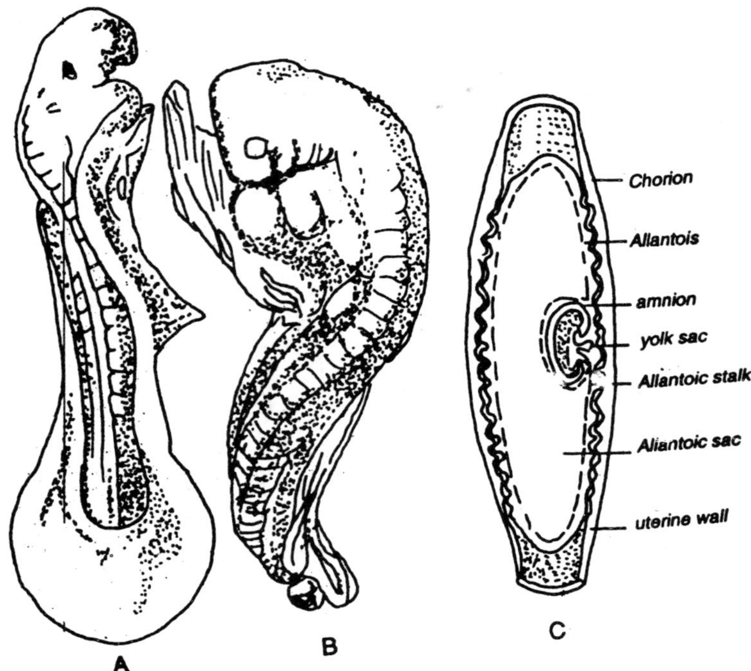

Figure 8.2 : Early pig embryos, in dorsal view. A, At sixteen days, with seventeen somites. B, At seventeen days, *approximately 4 mm. long, with about full number of somites. C, Hemisection of uterus, showing fetal membranes and placental relations.*

External Form

The *head,* which is relatively large on account of the dominance of the brain, makes a right-angled bend at the *cephalic flexure.* On the under surface of the head are the *olfactory pits,* now drawn into elongate grooves and bounded *by lateral* and *median nasal processes.* The lens of the *eye is* prominent as it lies beneath the ectoderm, surrounded by the optic cup. At the sides of the head are four *branchial arches,* separated by three *branchial grooves.* The first branchial arch of each side forks ventrally into two parts. The smaller *maxillary processes* show signs of fusing with the median nasal processes to form the upper jaw, while the larger *mandibular processes* have united already into the lower jaw. Next caudad is the prominent second, or hyoid arch *Small* tubercles, which will combine into the *auricle* of the external ear, bound the first branchial groove; the groove itself will become the *external acoustic meatus*. The third branchial arch is still visible in the future neck region, but the fourth arch has sunk into the *cervical* sinus; both disappear at a slightly later stage.

At the *cervical flexure* the head is bent at right angles to the body, thus bringing the ventral surface to the head close to the trunk; it is probably owing to this flexure that the third and fourth branchial arches buckle inward to give rise to the *cervical sinus*. Along its dorsal surface the trunk curves convexly, but this feature is not so prominent as at 6 mm. The reduction in trunk curvature results from the increased size of the heart, liver and mesonephroi. These organs are plainly indicated through the translucent body wall, while the position of the septum transversum may be noted between the heart and the liver. The *limb buds* are growing rapidly; the upper limbs are at a level between heart and liver. The *umbilical cord* is relatively large; it attaches at the lower end of the trunk. Dorsally the *somites* occur in serial order; toward the tail they become progressively smaller. Paralleling them and extending in a curve between the bases of the limb buds is the *mammary ridge;* on this thickened band of ectoderm will differentiate the mammary glands. The *tail* is long and tapering. Between its base and the umbilical cord is the genital tubercle.

Nervous System

Brain and Spinal Cord. Five distinct regions of the brain can be distinguished : (1) The *telencephalon* exhibits its rounded lateral outgrowths, the *cerebral hemispheres*. Their cavities, the *lateral ventricles*, communicate by interventricular foramina with the third ventricle. (2) T*he diencephalon* shows a laterally flattened cavity, the *third ventricle*. From the ventrolateral side of the diencephalon pass off the optic stalks, while an evagination of the midventral wall (the infundibulum) will produce the neural lobe of the *hypophysis. (3)* The *mesencephalon* never subdivides, and its cavity becomes the *cerebral aqueduct* leading caudad into the fourth ventricle. (4) The *metencephalon is* separated from the mesencephalon by a constriction, the *isthmus*. Dorsolaterally it becomes the *cerebellum,* ventrally the *pons*. *(5)* The elongate *myelencephalon is* roofed over by a thin and nonnervous ependymal layer. Its ventrolateral wall is thickened and still gives internal indication of the *neuromeres*. The cavity of the metencephalon and myelencephalon is the *fourth ventricle.*

The *spinal cord* begins without specific demarcation and extends into the tapering tail. Just beneath the hind-brain and spinal cord lies the *notochord.*

Cranial Nerves

Of the twelve pairs of cranial nerves, all but the olfactory and abducent are represented in where they occur in the order listed : (1) The *olfactory nerve is* not grossly demonstrable at this stage. (2) Fibers of the *optic nerve* are growing brainward within the optic stalk, cut through in this illustration. (3) The *oculomotor nerve,* a motor nerve to four of the eye muscles, takes origin from the ventrolateral wall of the mesencephalon and passes downward between the two parts of the bent brain. (4) The *trochlear nerve,* motor and destined for the superior oblique muscle of the eye, really arises from the ventral wall of the mesencephalon but emerges dorsally at the isthmus. The next eight pairs of nerves pass off from the rhombencephalon; four of these are rostral to the otocyst in the metencephalon and four lie caudally in the myelencephalon. (5) The *trigeminal nerve is* conspicuous because of its large *semilunar ganglion* and three branches (ophthalmic, maxillary and mandibular rami) which carry motor impulses to the jaw muscles and bring sensory impulses from the head. (6) The *abducent nerve* originates from the ventral brain wall and passes to the eye, where it will innervate the external rectus muscle. (7) The *facial nerve is* mixed, sensory and motor; it bears the *geniculate ganglion* and divides into chorda tympani, facial and superficial petrosal rami in the order named; most of the nerve has to do with the motor innervation of the face, whereas the sensory supply goes to the tongue. (8) The *acoustic nerve* arises just rostral to the otocyst; it bears the *acoustic ganglion* which will send sensory fibers to the internal ear. Caudal to the otocyst is the *glossopharyngeal nerve,* showing a proximal *superior* and a more distal *petrosal ganglion;* its sensory and motor fibers innervate both tongue and pharynx. (10) The *vagus nerve* is mixed in function and has both *a jugular* and a *nodose ganglion;* its fibers innervate chiefly the viscera. (11) The *accessory nerve* has motor fibers which take origin both from the lateral wall of the myelencephalon and from the spinal cord as far caudad as the sixth cervical ganglion; an internal branch accompanies the vagus, while the external branch is distributed to the sterno-mastoid and trapezius muscles. (12) The *hypoglossal nerve* arises by five or *six* rootlets from the ventral wall of the myelencephalon; it is purely motor and supplies the muscles of the tongue.

It should be noted that the fifth, seventh, ninth and tenth cranial nerves pass into the four branchial arches in the order named. This primitive relation is maintained in the adult when the nerves innervate the derivatives of these arches.

A nodular chain of ganglion cells extends caudad from the jugular ganglion of the vagus. These have usually been interpreted *as accessory vagus* ganglia. They may, however," be continuous with *Froriep's ganglion* which sends sensory fibers to the *n.* hypoglossus. In pig embryos of 15 mm. this chain is frequently divided into four or five ganglionic masses of which occasionally two or three (including Froriep's ganglion) contribute fibers to the root fascicles *of* the hypoglossal nerve.

Spinal Nerves

Each nerve has a single *spinal ganglion,* from which the sensory *dorsal root* fibers are developed. The motor fibers take origin from the ventral cells of the neural tube; they form the *ventral roots* which join the dorsal roots in the common *nerve trunk.* In the region of the upper and lower limb buds the spinal nerves unite and give rise respectively to the *branchial* and *lumbosacral plexuses.*

Autonomic System

Paired masses of ganglion cells, located dorsolateral to the descending aorta, are the forerunners of this division of the nervous system. These segmental masses are *autonomic ganglia* and, as ganglionated chains, become the *sympathetic trunks*. Fibers pass from the spinal cord to the ganglia and these fibers comprise a branch of the spinal nerve known as its *communicating ramus* (r. communicans).

Sense Organs

The *olfactory pits* are deep fossae, flanked by the median and lateral nasal processes. The stalked *optic cups* are prominent and the *lens vesicles* have detached from the ectoderm. Each *otocyst is* a compressed oval vesicle with a tubular *endolymph duct* growing from its medial side.

Digestive and Respiratory Systems

Mouth and Pharynx. The pharyngeal membrane disappeared at a considerably earlier stage and the *stomodeum,* or ectodermal mouth, then becomes continuous with the pharynx. From the dorsal wall of the stomodeum, *Rathke's pouch* (epithelial hypophysis) extends as a long, stalked sac which forks at its end near the primordium of the neural lobe. The floor of the mouth and pharynx is occupied by the *tongue* and *epiglottis.* From the mandibular arches arise paired *lateral swellings* that become the body of the tongue. Lying between these thickenings is the transient *tuberculum impar.* The thyroglossal duct, which formerly opened just caudal to the tuberculum impar, is already obliterated; the *thyroid gland* itself, composed of branching epithelial cords, is now located in the midplane between the second and third branchial arches. A median ridge, named the *copula,* unites the second arches and represents the primitive root of the tongue; it connects the tuberculum impar with the epiglottis which develops from the bases of the third and fourth branchial arches. Oh each side of the slit-like *glottis is* an *arytenoid fold of* the larynx.

The pharynx is flattened dorsoventrally; it is broad at the oral end. Opposite the third branchial arch the pharynx bends sharply in conformity with the cervical flexure. The paired *pharyngeal pouches* are large, and each bears a dorsal and a ventral wing. The first pouch on each side persists as the *auditory tube* and *tympanic cavity;* the 'closing plate' between it and the first branchial groove forms the *tympanic membrane,* while the ectodermal groove becomes the *external acoustic meatus.* The second pouches are destined largely to disappear; about each develops *a palatine tonsil.* The dorsal wing of each tubular third pouch forms a *parathyroid gland;* the ventral wings differentiate into the *thymus.* The fourth pouches are smaller; their dorsal wings give rise to another pair of parathyroids, while the ventral wings are rudimentary. A tubular outgrowth, just caudal to each fourth pouch, is sometimes regarded as a fifth phayngeal pouch; it forms an *ultimobranchial body.*

Larynx, Trachea and Lungs. The *larynx* and *epiglottis* are appearing and the *trachea is* a definite tube. Terminally the trachea bifurcates into *primary bronchi;* each of these has already divided again into secondary bronchial buds which indicate the two lobes of the *left lung* and the middle and lower lobes of the *right lung.* From the right side of the trachea itself appears another bud which, in the, pig, represents the upper lobe of the right lung.

Esophagus and Stomach. The *esophagus* extends as a narrow tube past the lungs, whereupon it dilates into the laterally flattened *stomach.* The entire stomach has rotated about its axis so that the original dorsal border, now the convex greater curvature, lies to the left and the primitive ventral border (lesser curvature) to the right. At this stage the rotation is incomplete.

Intestine

The pyloric end of the stomach opens into the *duodenum* which also shows the effect of stomach rotation; the stem of the *hepatic diverticulum,* originating from it ventrally, now lies to the right. The diverticulum itself. has differentiated into various things. From its tip has become the four-lobed *liver,* filling in the space between the heart, stomach and duodenum. One of the several ducts now connecting the liver with the parent diverticulum will persist as the *hepatic duct.* The main stem of the diverticulum is the *common bile duct,* while a side sacculation is the cystic duct and gall bladder. The *ventral pancreas* springs from the common bile duct near its point of origin. It is directed dorsad and caudad, to the right of the duodenum. The *dorsal* pancreas arises a little more caudally (man, cranially) from the dorsal wall of the duodenum; its, larger, lobulated body grows dorsad and cephalad. The two glands will interlock into a single organ; it the pig it is the duct of the dorsal pancreas that persists as the functional duct.

Beyond the duodenum the intestine is thrown into a loop, which extends well into the umbilical cord and connects with the *yolk stalk* there. Owing to rotation in the entire loop, the cranial limb of the intestine lies to the right, the caudal limb to the left. The small intestine (jejunum and ileum) extends as far as a slight enlargement on the caudal limb of the loop. This is *caecum* which marks the beginning of the *large intestine* (colon and rectum). The cloaca is now subdividing into the *rectum* and *urogenital sinus.*

Coelom and Mesenteries. The coelom is a continuous, communicating system which includes the single *pericardial* and *peritoneal cavities,* still connected by paired *pleural canals.* Between the heart and liver is a prominent partition, the *septum transversum;* the liver is broadly fused to this septum which will comprise much of the diaphragm. The double sheet of splanchnic mesoderm that serves as the primitive dorsal mesentery receives special names at different levels. Where it suspends the stomach it is known as the *mesogastrium,* or *greater omentum.* Then in order come the *mesoduodenum, mesentery* proper of the small intestine, *mesocolon* and *mesorectum.* The mesentery proper and some of the mesocolon follow the intestinal loop out into the umbilical cord. The ventral mesentery is limited in extent. It persists as the *lesser omentum* between stomach and liver, encloses the liver, and continues as the *falciform ligament* between the liver and ventral body wall. A saccular recess, between the caval mesentery and liver on the right and the stomach and its mesenteries on the left, is the *vestibule* and *omental bursa.* It opens through a narrowed *epiploci foramen.*

Urogenital System. The *mesonephroi* are large and complex in the pig. Along the middle of their ventromesial surfaces *genital ridges* have become prominent. In a ventral dissection the course of the *mesonephric ducts* can be traced along the ventral margins of the mesonephroi and into the *urogenital sinus*. The allantois is a conspicuous, stalked sac which communicates with the ventral part of the urogenital sinus.

The *metanephroi,* or permanent kidneys, lie far caudad between the roots of the umbilical arteries. At the present stage each consists of a tubular epithelial portion, surrounded by a mass of condensed mesenchyme. The epithelial tube has budded off the mesonephric duct, near its ending; proximally there is a slender duct, the *urter,* while a distal dilatation is the *renal pelvis*. From the pelvis grow out later the calyces and collecting tubules of the kidney. Encasing the pelvic primordium is a layer of condensed mesenchyme, derived from the lower nephrotomes and destined to differentiate into the secretory tubules, or *nephrons*.

Heart. This organ lies within the pericardial cavity. There are two *atria* and two thicker-walled *ventricles*. In addition, a small chamber, the *sinus venosus,* receives all the blood returned to the heart and directs it into the right atrium while the *bulbus cord is* still serves as a common arterial outlet.

The entrance from the sinus venosus into the right atrium is a nearly sagittal slit, guarded by right and left *valves of the sinus venosus (B)*. Dorsally the two valves join and continue a short distance as the temporary *septum spurium (A)*. Somewhat later the sinus largely loses its identity by merging with the right atrium, although its middle part does persist as the *coronary sinus*. The dorsal wall of the left atrium is receiving a single *pulmonary vein.* The two *atria* are incompletely partitioned by the *septum primum* which contains an opening, the *interatrial foramen.* On the right side of this partition a second sickle-like fold, the *septum secundum, is* forming. It also becomes an incomplete septum which bears an opening, known as the *foramen ovale.* After birth these two septa, together with the left valve of the sinus venosus, will fuse to complete the final atrial septum.

In a slightly younger embryo the atria and ventricles communicated through a common canal, bounded by two thickenings named *endocardial cushions.* At the present stage the two cushions have joined midway, have received the septum primum, and now subdivided this passage into two *atrioventricular canals.* About the right canal the endocardium is already undermined and in the process of forming the *tricuspid valve (b);* similarly, on the left is the developing *bicuspid valve.* The two *ventricles* are separated by *a ventricular septum;* it is complete except for, the *interventricular foramen* which connects the left ventricle with the bulbar part of the right (AB). The *bulbus cord is* separates distally into *ascending aorta* and *pulmonary artery,* but proximally it is still undivided (b). The ventricular walls are thick and spongy, forming a meshwork of muscular trabeculae separated by sinusoids. Until later, when coronary vessels are developed, the heart receives all its nourishment from the blood circulating in the sinusoids.

Vascular System. Arteries. The aortic-arch system is still represented, although somewhat modified and in process of transforming into its permanent derivatives. The first two pairs of arches have disappeared. The *ascending aorta* continues into the third and fourth pairs of arches. The third pair and the extensions of the dorsal aortae into the head constitute continuous channels, to be known as the *internal carotid arteries.* Near their bases arise the *external carotid arteries,* which extend into the region of the lower jaw. The fourth aortic arch is largest, and on the left side it will form the permanent *arch of the aorta.* The sixth (fifth?) aortic arches connect with the pulmonary trunk, and from them small *pulmonary arteries* pass to the lungs; the left arch continues until birth as the *duct us arteriosus.*

The paired *dorsal aortae* unite opposite the eighth somites and continue caudad as the median *descending aorta.* The aorta shows dorsal, lateral and ventral branches. The dorsal branches pass upward between the somites, and accordingly can be called *intersegmental arteries.* From the seventh pair, which is located just where the dorsal aortae combine, the *subclavian arteries* pass off to the upper limb buds, and *vertebral arteries* run cephalad into the' head. The latter vessels are formed by longitudinal anastomoses between the first seven pairs on each side, after which the stems of the first six atrophy. Under the brain the vertebrals are continuous with the unpaired *basilar artery;* the latter connects with the internal carotids beneath the diencephalon. Lateral branches of the descending aorta supply the mesonephroi and genital ridges. Ventral branches form the *coeliac artery* to the stomach region, the *superior mesenteric artery* (primitive vitelline) to the small intestine and the *inferior mesenteric artery* to the large intestine. The *umbilical arteries* (to the allantois and placenta) belong in this ventral series, but they now arise laterally from secondary trunks which persist as the *common iliacs*. Beyond this point the aorta narrows into the diminutive caudal *artery* extending into the tail.

Veins. Three sets of plexuses, which are the forerunners of the *dural sinuses,* occur alongside the brain. They drain into the *anterior cardinal veins, now* becoming the *internal jugular veins.* After receiving the newer *external jugular veins from* the mandibular region, the anterior cardinals open into the *common cardinal veins* (ducts of Cuvier). The latter vessels also receive the *sub-clavian veins* from the upper limb-buds and the *posterior cardinals* from the lower body. They empty into the sinus venosus. The *posterior cardinal veins* are the oldest veins caudal to the level of the heart. They course dorsal to the mesonephroi and drain the mesonephric sinusoids. However, the posterior cardinal veins are already beginning to decline, and midway along their lengths an interruption occurs; for this reason only the cranial halvesco-mmunicate with the common cardinal stems.

Considerable diversion of blood from the posterior cardinal veins has been brought about by the development of *subcardinal veins* along the ventromesial surfaces of the mesonephroL These vessels arose as longitudinal channels in a mesonephric plexus that was originally tributary to the posterior cardinal veins. Connections between the post- and subcardinal systems of each side still exist, while the two subcardinals also communicate by a prominent anastomsis across the midplane of the body. Their drainage is now shifting into the just organizing inferior vena cava. A fairly prominent *ventral vein* of the mesonephros follows the ventral border of this organ, but it soon disappears.

The *inferior versa cava is* becoming established at this stage. It has a compound origin, some of which is now identifiable. In the mesonephric region the larger right subcardinal is an important component. More cephalad a vein has developed in a specialized portion (caval mesentery) of the mesogastrium. This vessel connects the subcardinal with the hepatic (vitelline) sinusoids). The blood-flow through the sinusoids is already consolidating into a definite channel, and this is the hepatic part of the inferior vena cava. It empties into the common hepatic vein (primitive right vitelline), which constitutes the stem of the vena cava.

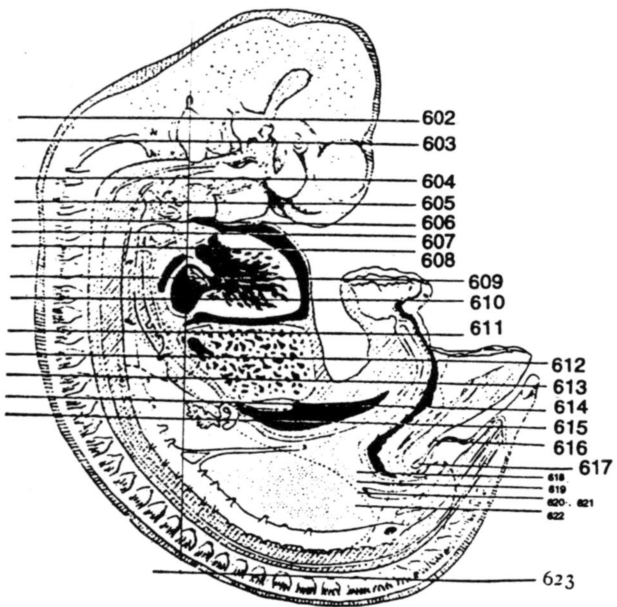

Figure 8.3 : Reconstruction of 10 mm. pig embryo.

The *umbilical veins* follow the allantoic stalk back from the placenta. In the umbilical cord they have merged into a common vessel, but they separate again on entering the embryo where they course in the ventrolateral body wall of each side to the level of the liver. The left umbilical is the larger of the two, and it alone persists in older fetuses. Cephalad of the liver the original stems that connected with the sinus venosus have disappeared, and umbilical blood is now routed through the liver in sinusoidal channels. Such an enlarged fetal passage, connecting the left umbilical with the inferior vena cava, is the important *ductus venosus*.

Distally the two *vitelline veins* are fused. Passing inward from the regressive yolk sac, they course cephalad of the intestinal loop. In the pancreas-region the left vein receives the *superior mesenteric vein* which is a new vessel arising in the mesentery of the intestinal loop. Above this junction a cross anastomosis and a continuation of the right vein make a new channel which is known as the *portal vein*. It gives off branches to the hepatic sinusoids, which have arisen much earlier from a breaking up of the vitelline veins in this region, and connects with the left umbilical vein within the liver. Beyond the sinusoids the vitelline vessels are retained as *hepatic veins* and the stem of the inferior *vena cava*.

TRANSVERSE SECTIONS OF A TEN M.M. PIG EMBRYO

The more important levels, are illustrated and described. These are useful for the identification of organs, but the student must interpret his sections with reference to the dissections and reconstructions. All sections are drawn from the cephalic surface; accordingly, the right side of the embryo is at .the reader's left.

Section through Cephalic Flexure

Since the head is flexed, the sections first encountered pass through the *mesencephalon* and *metencephalon*. At a slightly lower level the metencephalon also becomes continuous with the thin-roofed *myelencephalon*, but presently the mid-brain gives way to the *diencephalon* as the brain becomes cut twice. Several important structures should be identified in the mesenchyme between these two portions of the brain. In the midplane, but nearer the metencephalon, is the single *basilar artery;* ventrolateral to the diencephalon are the paired *internal carotids*. These three vessels unite at the location of the future *arterial circle* (of Willis). About halfway between the midplane and the lateral wall appear branches of the *anterior cardinal veins* and the *oculomotor* and *trochlear nerves*. Of the two nerves, the trochlear is smaller and slightly more lateral in position; in some series it extends only a slight distance.

Section through Infundibulum, Semilunar Ganglia and Otocysts

The brain is sectioned twice. At the bottom of the section is the *diencephalon,* cut transversely; its cavity is the *third ventricle.* Mid ventrally the diencephalon gives off the *infundibulum* which furnishes the neural lobe of the *hypohysis.* The *metencephalon* and *myelencephalon* are sectioned frontally; there is no clear demarcation between the two. Their walls bear the prominent scalloping of the *neuromeres,* while the common cavity is the *fourth ventricle.* The wall of the entire neural tube now shows a differentiation into three layers : (1) an inner *ependymal layer,* densely cellular, next the central canal; (2) a middle *mantle layer,* of nerve cells and fibers; and (3) an outer *marginal layer,* chiefly fibrous. A thin, vascular layer surrounds the brain wall as the primitive *pia mater.*

The interval between the two portion of the brain contains several structures, sectioned transversely. Next the metencephalon is the unpaired *basilar artery;* ventrolateral to the diencephalon are the paired *internal carotid arteries.* Near the latter the *oculomotor nerves.* In this embryo the trochlear nerves had not grown down to this level. On the left side is a part of the ophthalmic branch of the *trigeminal nerve.* Tributaries of the *anterior cardinal veins* occur, the largest rostral to the semilunar ganglia. This is a portion of the *cavernous sinus,* while the stem of the *middle dural plexus is* just caudal to the semilunar gaglion; alongside the diencephalon the *lateral sinus* is cut.

Near the beginning of the hind-brain are the large *semilunar gangila;* from their medial sides nerve fibers of the *trigeminal nerves* join the brain wall. This ganglion, situated at the pontine flexure of the metencephalon, constitute one of the most important landmarks of the embryonic head. Slightly caudad lie the *facial nerves;* the left is cut as it leaves the brain wall. Midway along each side of the hind-brain will be seen the apex of an *otocyst,* and mesial to it, the *endolymph duct;* on the left side the latter is cut near its origin from the otocyst.

Section through Cerebral Hemispheres and Eyes

This level shows some important new features. The *dience-*

phalon is now continuous with the *telencephalon.* The latter consists of a mesial region which has evaginated paired *cerebral hemispheres;* their cavities, the *lateral ventricles,* connect through the *interve-ntricular foramina* with the *third ventricle* of the diencephalon. Close to the ventral wall of the diencephalon is a section of the epithelial lobe of the hypophysis *(Rathke's pouch),* near which are the *internal carotid arteries.* Lateral to to the diencephalon are the *optic cups,* sectioned just caudal to their stalks. The double wall of the optic cup comprises the *retina;* the thin outer layer is the pigmented epithelium, the inner and thicker coat is the nervous layer. The *lens is* now a closed vesicle, distinct from the overlying *corneal ectoderm.*

The irregular vascular spaces are tributaries of the *anterior cardinal veins.* The largest space is the *cavernous sinus, in* the vicinity of the fifth nerve. The upper half of the section contains portions of the *posterior dural plexus,* while the two small vessels between the cerebral hemispheres at the bottom of the section represent the *superior longitudinal sinus.*

By working above and below the present level all the cranial nerves and ganglia, as well as the central connections and peripheral courses of these nerves, will be observed. In Fig. 3., transverse sections of the maxillary and mandibular branches of the *trigeminal nerve* are seen, while the *abducent nerve is* sectioned longitudinally as it passes from the under surface of the hind-brain toward the eyes. On the rostral side of the otocyst occur the *geniculate ganglion* of the *facial nerve* and the *acoustic ganglia* of the *acoustic nerve.* The *otocyst* is a sharply defined, epithelial sac which lies at the junction of the metencephalon and myelencephalon and makes a convenient landmark in identifying ganglia and nerves. Caudal to the otocyst, the *glossopharyngeal nerve* and the *jugular ganglion* of the *vagus nerve* are cut transversely while the trunk of the *accessory nerve is* sectioned lengthwise as it curves forward from the level of the spinal cord.

Section through Mouth, Tongue and First and Second Pouches

The tip of the head, with parts of the *telencephalon* and *olfactory pits, is* now separate from the rest of the section. Since the level last described, Rathke's pouch has opened into the ectodermal *stomodeum* between the jaws; the present section is at the actual *oral opening,* bounded by the *maxillary* and *mandibular processes* of the *first branchial arches.* At a slightly lower level, the mandibular processes merge in *lateral lingual swellings* which will become the body of the *tongue.* With the disappearance of the pharyngeal membrane, the stomodeum and entodermal mouth cavity have become continuous. The *pharynx* shows ventral portions of the *first* and *second pharyngeal pouches,* destined to be utilized as auditory tubes and tonsillar fossae, respectively. Opposite the first pouch, externally, is the *first branchial groove,* or future external acoustic meatus. A shaving has been cut from the *tuberculum impar* of the *tongue* as it rises above the floor of the pharynx; the upper (caudal) end traces into connection with the second arches which unite as the *copula.* The *facial nerves* of the *second branchial arches* are cut across, but since the previous level the *trigeminal nerves* have ended in the maxillary and mandibular processes of the first arches.

The *mylcncephalon is* sectioned close to its continuation into the spinal cord; *Froriep's ganglion* and some of the *accessory nerve* are included. Between the myelencephalon and the pharynx are seen on each side the several rootlets of the *hypoglossal nerve,* fibers of the *vagus* and *accessory nerves,* and the *petrosal ganglion of* the *glossopharyngeal nerve* which stands in relation to the third arch. Mesial to the petrosal ganglia are the *dorsal aortae,* and lateral to the vagi are the *anterior cardinal veins.* In the midplane is a bit of the *notochord,* cut lengthwise. The *basilar artery* still **lies beneath** the myelencephalon, but a short distance caudad it is replaced by the paired *vertebral* arteries.

Section through Third Pouches and Olfactory Pits

The tip of the head is now small and includes on either side the open *olfactory pits,* lined with thickened ectodermal epithelium. Each pit is bordered by a *lateral* and a *median nasal process,* which assist in the formation of the nose and upper jaw. In preceding sections the epithelium within the pit forms a superficial plate that is the *oro-nasal membrane;* it ruptures later and produces a *primitive choana.* The first three pairs of *branchial arches* show; the first is fused as 'the *mandible,* and the third is slightly sunken in the *cervical sinus.* The dorsal wings of the *third pharyngeal pouches* extend toward the ectoderm of the *third branchial grooves;* attached to these wings are prominent *parathyroid* primordia. The ventral wings are large, epithelial sacs that can be followed into succeeding sections; one is shown here as a separate drawing. They give rise to the *thymus.* The floor of the pharynx is sectioned through the *epiglottis.*

Ventral to the pharynx are portions of the *third aortic arches* (internal carotids) and the solid cords of the *thyroid gland. (External carotid artieries* arise at a slightly higher level, close to the ventral origins of the third acrtic arches; they can be traced for a variable distance into the substance of the lower jaw). Beneath the thyroid and in the mandible are portions of the *external jugular veins.* Dorsally the section passes through the *spinal cord* and the first pair of *cervical ganglia.* Between the spinal cord and pharynx, laterally, are the *anterior cardinal veins,* the *hypoglossal nerves* and the *nodose ganglia* of the *vagi.* Lateral to each ganglion is the external branch of the *accessory nerve* and mesial to the ganglia are the small *dorsal aortae.* The *notchord is* conspicuous in the midplane.

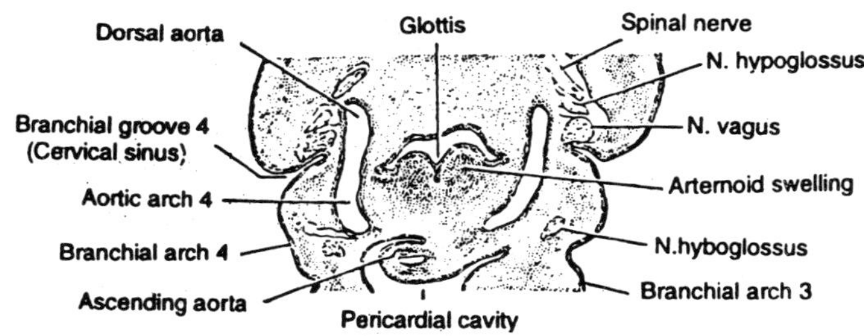

Figure 8.4 : Transverse section through glottis and ourth aortic arches of 10 mm. pig embryo.

Section through Glottis and Fourth Aortic Arches

At this level the first three branchial arches have been passed and the cephalic border of the heart is coming into view. The illustration includes only the region of the *fourth bronchial arches,* and the fourth pair of *aortic arches* which course through them. The left aortic arch will become the permanent *arch of the aorta,* the right the stem of the right *subclavian artery.* The *ascending aorta* connects with these aortic arches a little cephalad in the series. The *fourth branchial groove* opens into the *cervical sinus* which is seen here. This duct extends for some distance before coming in contact with the fourth pharyngeal pouch. The section cuts across the *pharynx,* the *glottis* (entrance to the larynx), and its bordering *arytenoid swellings.* In addition to a spinal nerve, section of the *vagus* and *hypoglossal nerves* are encountered.

Section through Fourth Pouches and Larynx. The head has been passed and the section is now dominated by the *heart,* lying within its *pericardial cavity.* The tips of the *atria* are sectioned as they project at the sides of the *bulbus cord is.* The bulbus is dividing into the aortic stem and the pulmonary trunk. The small section of the *ascending aorta* traces cephalad into the third and fourth aortic arches and caudad into the ventricle. The *pulmonary trunk is* cut twice; its distal portion can be followed caudad into connection with the sixth aortic arches.

The crescentic *pharynx is* continued laterad as the small *fourth pharyngeal pouches.* Each gives origin to a dorsal wing *(parathyroid),* encountered a few sections cephalad in the series and shown here as a separate drawing. At the level of the main section the pharynx is also continuous with a saccular *ultimobranchial body.* From the midventral wall of the pharynx arises the solid epithelial plate of the *larynx.* A section of the vagus nerve is located between the dorsal aorta and the *anterior cardinal vein of* each side. Ventral to the anterior cardinals (soon to be called the *internal jugular veins)* are small *external jugular veins.* The left *dorsal aorta is* larger than the right in anticipation of its conversion into the permanent descending aorta of this level.

Section through Pulmonary ARches and Bulbus Cordis

The heart is little changed from the last level. However, the *ascending aorta* communicates with the *bulbus,* while the latter shows on each side the thickened internal *ridges* that will progressively meet and continue the separation of the aortic and pulmonary trunks. The *sixth aortic arches* connect the dorsal aortae with the main *pulmonary trunk,* whose origin was traced in the preceding section. On the left side of the embryo the arch is complete; it represents the *ductus arteriosus,* which remains patent until birth. From these pulmonary arches small *pulmonary arteries* may be traced caudad in the series toward the lungs. The *esophagus* is now separate from the *trachea;* both are cut through their extreme cephalic ends. Ventrolateral to the spinal cord are somewhat diffuse *myotomes,* while *sclerotome* masses surround the notochord. The vagus *nerves* are prominent. At about this level the external jugular veins join the *anterior cardinals.*

Section through Heart and Interatrial Foramen

Only the heart is figured, taken from a total section much like. All four chambers are shown. The *Atria* are partially separated by the *septum primum,* which is incomplete because of the *interatrial foramen;* this foramen will remain open until birth. A septum secundum is not seen satisfactorily in embryos of this size. Each atrium communicates with the ventricle of the same side through an *atrio-ventricular canal.* Between these openings is the fused portion of the *endocardial cushions.* At an earlier stage these cushions were double, but they have fused midway and thus divide the originally single canal into two; they will also help in the formation of the *bicuspid* and *tricuspid valves.* The atria are marked off externally from the ventricles by the *coronary sulcus.* Between the two ventricles is the *ventricular septum; a* little higher in the series of sections the thin region (above leader) gives way to a temporary *interventricular foramen.* The ventricular walls are thick and spongy, forming a network of muscular *trabeculae* surrounded by blood spaces, or sinusoids. This muscular layer constitutes the *myocardium.* It is lined by an endothelial layer, the *endocardium,* while the entire heart is surrounded by a layer of mesothelium'the *epicardium,* or visceral pericardium.. The latter sac is continuous with the parietal pericardium, which lines the body wall in the region of the heart.

Section through Common Cardinals and Sinus Venosus

The section is also marked by the large *heart* and the bases of the upper *limb buds*. Dorsal to the atria are the *common cardinal veins*. The right vein empties into the *sinus venosus*; the left crosses the midplane and connects with the sinus at a lower level. just above the plane of. this section the right common cardinal has received the right *subclavian vein* from the limb bud; the left subclavian ıs still separate. The sinus venosus drains_ into the right atrium through a slit-like opening in the dorsal and caudal atrial wall. The opening is guarded by the right and left *valves of the sinus venosus,* both of which project into the atrium. The *septum primum* appears like a complete membrane at this level, which is a little caudal to the interatrial foramen and the atrioventricular canals. The septum joins the fused *endocardial cushions,* as does the *ventricular septum* from below.

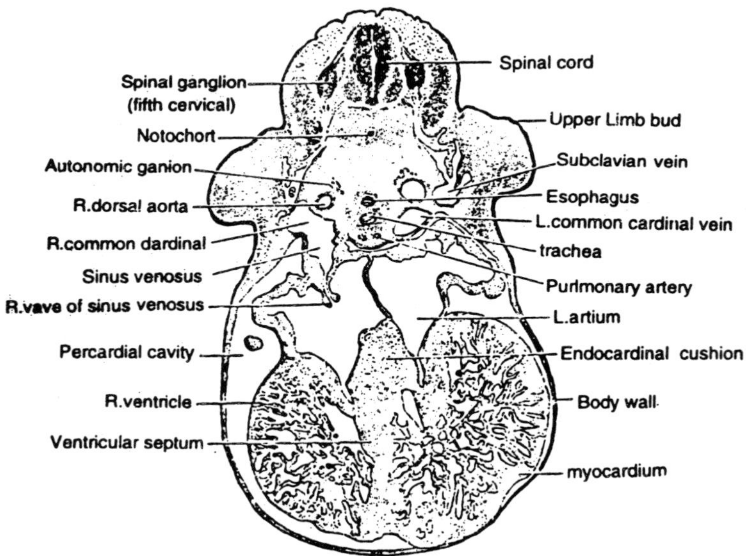

Figure 8.5. Transverse section through common cardinals and sinus venosus of 10 mm. pig embryo.

The *esophagus* and *trachea* are tubular. Around the epithelium of both are condensations of mesenchyme from which their fibrous and muscular layers are to be differentiated; laterally in this mass lie the *vagus nerves,* unlabeled in the illustration. Ventral to the trachea are the *pulmonary arteries*. The left *dorsal aorta* is large and is here continuing the arch of the aorta caudad; the right dorsal aorta at this level forms a part of the right subclavian artery. dorsolateral to the aortae are *autonomic ganglia.* The condensation of mesenchyme about the notochord foreshadows a future *vertebra.*

Section through Branchial Plexus and Tracheal Bifurcation

Other distinctive features are the presence of upper *limb buds* and the *liver.* The seventh pair of cervical *spinal nerves is* cut lengthwise in diagrammatic fashion. The spindle-shaped ganglion is associated with the *dorsal root* whose fibers grow out from its cells; the fibers of the *ventral root* arise from the middle, cellular layer of the cord and join the dorsal root in the common *nerve trunk.* On the right side, a short *dorsal ramus* unites with similar rami of several other nerves to form the *brachial plexus, a* part of which is seen. Adjacent to the esophagus are the cut *vagus nerves,* which follow this tube laterally.

The *descending aorta* shows its manner of origin from paired vessels. From the seventh pair of *intersegmental arteries,* which arise dorsally from it, the *subclavian arteries* are given off two sections caudad in the series. Traced cephalad these seventh intersegmentals become continuous with the *vertebral arteries.* The latter lie mesial to the stem portions of the spinal nerves. In some embryos they are imperfectly developed at this stage. Lateral to the aorta are the *posterior cardinal veins,* easily traced to the common cardinals of the previous figure.

Beneath the *esophagus* the trachea branches into short *primary bronchi.* These continue laterad into buds which represent the upper left and middle right lobes of the *lungs.* The upper right bud (unpaired) comes off the trachea a little cephalad in the series of the paired lower lobes are more caudad. Crescentic *pleural cavities* bound the bulging pulmonary tissue laterally.

On the embryo's left this cavity is separated from the peritoneal cavity by *a pleuro-peritoneal membrane,* and here the parietal and visceral pleura assume typical relations. The broad mesial mass of mesenchyme containing the esophagus and lungs is the *mediastinum.* Ventral to the bifurcating trachea are sections of the *pulmonary veins,* which can be traced into the left atrium at a slightly higher level. The liver, with its close network of trabeculae and sinusoids, is large and nearly fills the *peritoneal cavity.* A little cephalad in the series its attachment to the septum transversum is seen better. In the present section the *inferior vena cava is* leaving the liver and entering the septum. This portion of the vena cava is, in reality, the *common hepatic vein* (stem of the primitive right vitelline).

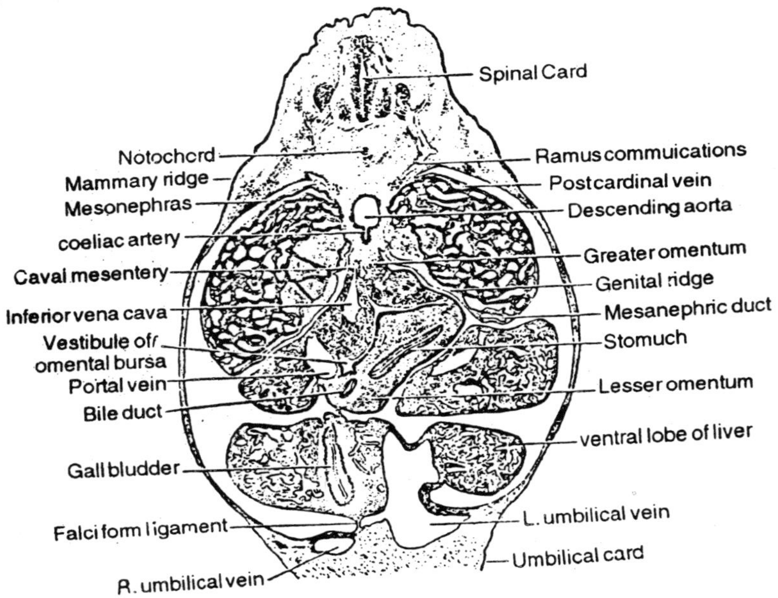

Figure 8.6 : Transverse section through pyloric stomach and gall bladder of 10 mm. pig embryo.

Section through Lungs, Upper Limbs and Autonomic Ganglia. The *limb buds* are' ectodermal sacs stuffed with dense, undifferentiated mesenchyme. Ectodermal thickening, at the tips, exert control over their progressive organization. Flanking the now circular *descending aorta* are the cranial ends of the *mesonephroi,* while above the aorta is a pair of *autonomic ganglia;* the *communicating rami* connecting the spinal nerve trunks do not show but can be seen in neighboring sections. The esophagus is just beginning to dilate into the *stomach,* and at this level the *vestibule* of the omental bursa appears as a crescentic slit to the right and below it. The mediastinum of higher levels has changed into a typical mesentery *of* the lower esophagus *(mesoesophagus)* and stomach *(mesogastrium,* or *omentim).* The *lungs* are sectioned through their paired lower lobes. Both *pleural cavities* still communicate freely with the peritoneal cavity. In the right dorsal lobe of the liver is located more of the intrahepatic portion of the *inferior vena vaca;* this particular segment is organizing from enlarged hepatic sinusoids. Near the midplane is the large *ductus venosus,* which traces into union with the vena cava a short distance cephalad. The *posterior cardinal veins* are coming into intimate relation with the *mesonephroi;* these temporary kidneys are out near their cranial ends.

Section through Stomach and Liver

It is *the stomach* and lobate *liver* that feature this level. The stomach has rotated partially, so that its original dorsal margin is now dropping to. a position on the embryo's left whereas the primitive ventral margin is raising correspondingly an the right. These margins are to be the *greater* and *lesser curvatures,* respectively: The stomach is attached dorsally by the *greater omentum;* ventrally the *lesser omentum* passes to the liver. This ventral mesentery splits into halves and is continued as a peritoneal reflection around the liver; the component layers then come together again as the *falciform ligament,* attaching the midventral border of the liver to the body wall. Both the body wall and the abdominal viscera are thus seen to be surfaced with a continuous sheet of mesothelium underlaid by mesenchyme; this serious investment is the *peritoneum,* in which a parietal and a visceral division is recognized.

The liver shows paired dorsal and ventral lobes. The right dorsal lobe is fused dorsally to the greater omentum. The connection forms the *caval mesentery,* in which courses the *inferior vena cava.* Between the attachments of the stomach and liver, and to the right of the stomach, is the *vestibule* of the omental bursa. Midventrally in the liver is the *ductus* venosus,sectioned just at the point where it receives the intraphepatic continuation of the

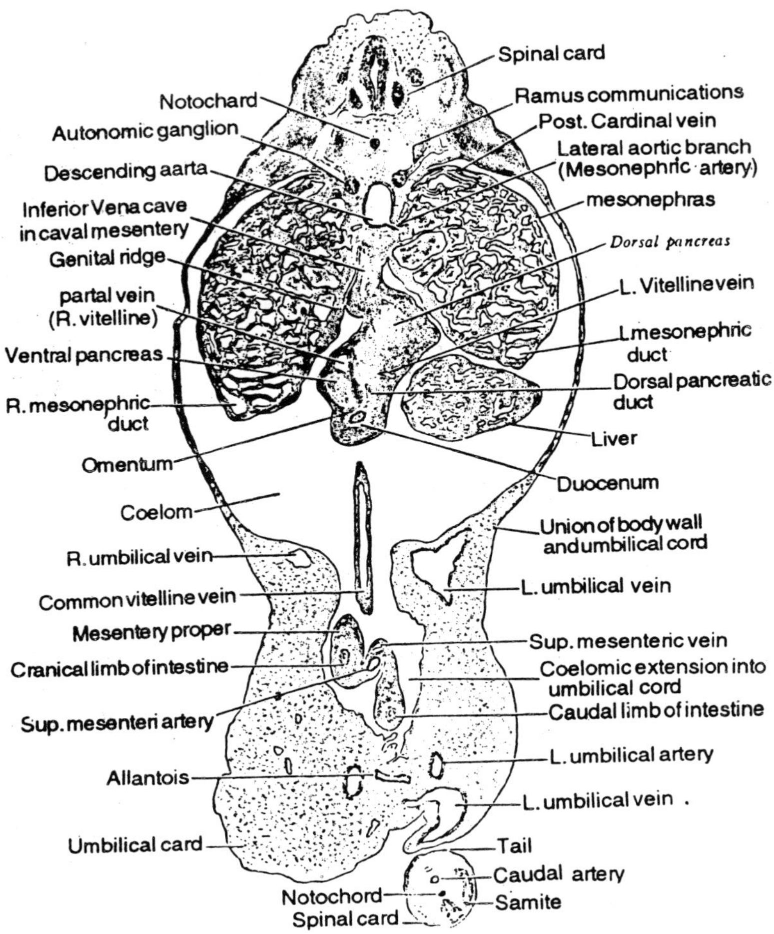

Figure 8.8 : Transverse section through pancreas, intestinal loop and tail of 10 mm. pig embryo.

left *umbilical vein* and a branch from the *portal vein.* The liver tissue is a complicated network of trabeculae and sinusoids; the component *liver cords* are composed of hepatic cells, surrounded by the endothelium of the sinusoids; red blood cells are developing here at this stage.

The *mesonephroi* are becoming prominent organs. Along their ventral margins course the *mesonephric ducts;* each shows a connection laterally with the terminal segment of a *collecting tubule.*

Section through Pyloric Stomach and Gall Bladder

The section grazes the third thoracic *ganglia,* passes slantingly through the pyloric end of the *stomach, then* cuts the *con non bile duct,* and finally passes lengthwise through the *gall bladder* superficially embedded in the substance of the *liver.* Within the distance of a few sections it is easy to demonstrate the continuity of these components parts of the digestive system. The *greater omentum* of the stomach is larger and more folded than in the previous illustration, and the *omental* bursa (lying horizontally over the stomach) is correspondingly expansive. It opens into a vertical space which is the *vestibule;* traced caudad a short distance the region of the vestibule, opens into the peritoneal cavity through the *epiploic foramen* of (winslow). Beneath the stomach is a caudal portion of the *lesser omentum.* The blood supply of the stomach-pancreas region comes from the *caeliac artery,* which is seen emerging as a ventral branch of the aorta.

The liver has nearly been left behind, and its dorsal and ventral sets of lobes are now separate. Associated with the liver are several veins. Dorsally is the *inferior vena cava,* about to leave the liver within a lip-like fold, the *caval mesentery.* Also in the right upper lobe is the *portal vein.* Ventrally, where the umbilical cord joins the body wall, are the *umbilical veins;* the larger left vessel is entering the left ventral lobe of the liver on its way to the ductus venosus. On each dorsolateral surface of the trunk is a thickened ectodermal ridge, which represents the *mammary ridge.* At the same horizontal level a *ramus communicans* passes from the left nerve trunk to an *autonomic ganglion.* The bottom of the figure includes a little of the insertion of the *umbilical cord* on the abdominal wall.

Section through Pancreas, Intestinal Loop and Tail. The bulging *mesonephroi* are conspicuous. *Mesonephric corpuscles,* with vascular *glomeruli* indenting them, are medial in position. The glomerull receive *mesonephric arteries,* arising as lateral branches from the aorta; one of these is shown. The *mesonephric tubules* are contorted and variously sectioned; they are lined with a cuboidal epithelium and empty into the *mesonephric duct* coursing along the ventral margin of the gland. On the mesial surface of each mesonephros the epithelium is thickened; from these *genital ridges* the sex glands will differentiate. The *posterior cardinal veins* lie on the dorsal surfaces of the mesonephroi, and transient *ventr: ! veins* of the mesonephroi course along the ventral borders. The *inferior vena cava* is a vertical slit in the caval mesentery. Traced caudad a short distance it joins the right subcardinal vein, which continues this important venous channel down the trunk. The *subcardinal veins* are prominent vessels,, seen well at a slightly lower level on the ventromesial surfaces of the *mesonephroi.* There they interconnect by a large anastomosis.

The *duodenum* lies within its dorsal mesentery *(mesoduodenum).* The duct of the lobulated *dorsal pancreas is* shown arising directly from the duodenal wall. More to the right is a section of the *ventral pancreas* which traces cephalad to its origin from the stem of the common bile duct. On each side of the dorsal pancreas are portions of the *vitelline veins,* the right at this level being a part of the *portal vein. A* few sections caudad this vessel can be followed across a transverse anastomosis to the left vitelline vein where the *superior mesenteric vein* from the mesentery attaches. Beyond the duodenum the vitelline veins have fused into a common vessel, cut lengthwise in the present section; it can be followed to the yolk sac, where the right and left components again separate.

The ventral body wall is continuous with the *umbilical cord.* This contains as extension of the embryonic *cvaelom,* and a portion of the *intestinal loop* within its *mesentery.* Between the two intestinal limbs are the *superior mesenteric artery* and *vein.* The former is a ventral branch of the aorta; the latter joins the portal vein. The *allantois is* flanked *by umbilical arteries,* while *umbilical veins* are cut in the cord and again as they enter the body wall. The tip of the recurved *tail* shows as a separate section; its structure is simple.

Section through Cloacal Membrane

To maintain the proper relations with sections already studied, this and succeeding sections through the curved, caudal region are shown dorsal side down. The caudal end of the embryo is small. Its laggard differentiation, in comparison to higher levels, is reflected in the less specialized, *spinal cord* and *somites.* the slender *tail gut* is cut across. Between the *notochord* and tail-gut is the continuation of the aorta, known here as the *caudal artery.* On each side of the latter lies the termination of a *posterior cardinal vein.* The ventral half of the section is featured by an epithelial plate that represents the solid *cloacal membrane.* The plane of section is such that the fusion of ectoderm with the entoderm of the plate is shown only at one end.

Section through Subdividing Cloaca. Tracing a short distance down the series from the previous level, the tail-gut joins the *cloaca* and the latter gains a cavity. Still farther, at the present level, the cloaca is separating into a dorsal *rectum* and a ventral *urogenital sinus.* (If should be understood that the recurvation of the tail-end of the embryo makes caudal progress in the sectioned series actually cephalad on this part of the embryo.

Section through Allantois, Urogenital Sinus and Rectum

This illustration (and the four that follow) include only the caudal, recurved part of the embryo. In the ventral body wall is seen the *allantoic stalk,* accompanied by the *umbilical arteries;* slightly cephalad in the series they all enter the umbilical cord. More dorsad are the crescentic *urogenital sinus* and the *rectum.* now separate. The caudalmost portion of the *caelom* tapers to an end between the two.

Section through Sterns of Mesonephric Ducts and Allantois

The *colon is* contained within a portion of the mesentery that is specifically named the *mseocolon.* In the body wall are tributaries of the *umbilical veins,* and mesial to them are the *umbilical arteries.* The *allaitcic stalk is* sectioned as it opens into the *urogenital sinus.* Dorsal to the sinus is a section of *rectum,* separated from the from the sinus by a crescentic prolongation of the *caelom.* The rectum has no mesentery. The hcrns of the urogenital sinus receive the *mesonephric ducts.*

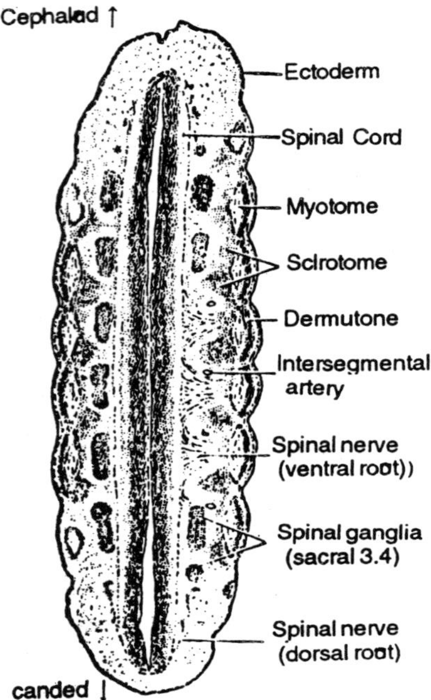

Figure 8.9 : Transverse section through curved back of 10 mm. pig embryo.

Section through Lower Limb Buds and Ureteric Stems

The section cuts through the middle of both lower *limb buds.* Like the upper set already studied, they consist of undifferentiated mesenchyme contained within an epithelial sac whose apex shows marked thickening. Mesial to the limb buds are the *umbilical arteries,* which in turn lie lateral to the *mesonephric ducts.* The left mesonephric duct is cut at just the proper plane to show the *ureter,* or duct of the mesonephric duct is cut at just the proper plane to show the *ureter,* or duct of the metanephros, being given off dorsally. The right ureter is sectioned transversely and appears as a separate tube. The *colon* is cut lengthwise; tracing cephalad in the series, it becomes continuous with the colon and rectum.

Section through Mesonephric Ducts and Ureters

Continuing down the series the ureters can be traced for some distance. In it the right *ureter* is seen. Also the right *mesonephric duct is* cut lengthwise (frontally) as it leaves the mesonephros on its way to connect with the urogenital sinus.

Section through Metanephroi

The ureters are found to terminate in the *metanephroi,* Figured here. Each of these kidney primordia consists of two parts. Internally there is a dilated expansion of the ureter that represents the *renal pelvis;* from it, first the *calyces* and then the system of *collecting tubules* will bud and grow. The periphery of the double primordium is a mass of condensed mesenchyme, derived from nearby nephrotomes; this tissue will differentiate into the *secretory tubules* of the kidney. *Iliac veins,* branches of the posterior cardinals, are passing into the limb buds.

Section through Curved Back

Owing to the lumber curvature, this section is actually frontal. The *spinal cord is* cut lengthwise. It is flanked by *spinal ganglia,* except midway on the left side where a slightly deeper plane includes several *spinal nerves* The ganglia include the third lumbar to the fourth sacral. The sometimes show spindle-shaped *myotomes* and more laterally located *'dermatomes.'* The medial side of each somite is a *selerotome;* this shows subdivision into a caudal denser and a cranial less dense half. Recombination of the dense half of one somite with the sparser half of the somite next caudad will produce a definitive *vertebra. Intersegmental arteries* appear between some of the sometimes on the left side.

ANATOMY OF AN EIGHTEEN MM. PIG EMBRYO

Most of the important organs are laid down in 10 mm. embryos. Older stages are chiefly instructive, therefore, to demonstrate the growth and differentiation of parts already present, rather than the introduction of new ones. Dissections show perfectly the form and relations of organs, their relative rates of growth and changes of position. Since the illustrations indicate better than descriptions the several structures and their states of development, only certain selected features will be mentioned.

External Form

The neck and back are much straighter than before, but the

ventral body is still highly convex. The head is relatively larger, the umbilical cord smaller. The sense organs are prominent, and the face, with snout and jaws, is plain. The branchial grooves and cervical sinus have disappeared from the neck. The limbs show indications of proximal and distal divisions, and the hand and foot are paddle-like Several mammary gland primordia occur along the mammary ridges, now locate, more ventrally. The genital tubercle has become a distinct phallus.

Lateral Dissection

The cerebral hemispheres are larger and the cerebellum is appearing. Beneath the cerebellum is the prominent pontine flexure of the brain, pointing ventrad. Nerves and ganglia show clearly; the brachial and lumbo-sacral plexuses, opposite the limbs, are noteworthy. The liver and lungs are relatively larger and more plainly lobed than before; the heart and mesonephroi are smaller.

Midsagittal Dissection

The corpus strain has developed in the floor of the cerebral hemisphere, a chorioid plexus invades the fourth ventricle, and neural (posterior) lobe of the hypophysis is growing into association with the detached Rathke's pouch. Sclerotomic primordia of vertebrae condense about the notochord. The viscera show only quantitative changes from the 10 mm. stage, but the urogenital sinus and rectum are now separate, as are the aorta and puimonary artery. The intestinal loop has rotated until the cranial and caudal limbs lie right and left, respectively. The caecum is conspicuous and a urinary bladder has developed between the allantois and urogenital sinus.

Ventral Dissections

The lungs, septum transversum, stomach, intestine and mesonephroi are the chief organs seen in the 15 mm. Of special interest are the beginnings of the Mullerian ducts.

ANATOMY OF A THIRTY-FINE MM. PIG EMBRYO

External Form

The embryo is straighter, slenderer and its ventral surface less protuberant. The head, with its prominent snout, is shaping like that of a lower mammal, and the neck becomes distinct. Digits have appeared on the elongate extremities. The umbilical cord and tail are losing rapidly in relative size.

Lateral Dissection

The spinal cord and brain are relatively smaller, but the latter is becoming highly specialized and folded. The cerebral hemispheres are large, and olfactory lobes extend forward from the rhinencephalon. The body of the embryo elongates faster than the spinal cord, so that the spinal nerves, at first directed at right angles, course obliquely in the lumbosacral region. Note especially how the viscera have 'receded' caudad and how the liver dominates the abdomen as the mesonephros loses prominence. The kidney is exceptional in that it shifts cephalad.

Midsagittal Section

New features of the brain are the olfactory lobes, the chorioid plexus of the third and lateral ventricles, the thalami, the epiphysis and the consolidated hypophysis. The primitive mouth cavity is now divided by the palatine folds into the upper nasal passages and lower oral cavity. Of the viscera, the distinct genital and suprarenal glands and the enlarged metanephros command attention, as does the coiling of the intestine. The ureters have acquired separate openings at the base of the bladder, and the urethra extends to the tip of the phallus.

Ventral Dissection

The chief new features are the markedly lobate lungs and the longer Mullerian ducts with expanded upper ends. The mesonephros is nearing its maximum absolute size.

9

Development of Mouse

The laboratory mouse has the longest history of genetic experimentation of any animal. Unlike *Drosophila,* developmental hypotheses and phenotypic tests of gene effects accompanied the studies of inheritance from the beginning of genetic analysis. In consequence, more is known about the developmental genetics of the mouse than that of any other vertebrate. The mouse as a model developmental system has a twofold interest. In the fist place, it may be taken as a representative vertebrate. Although the early development of some vertebrates such as amphibians appears very different from that of mammals, these differences may be less significant than was once thought. Furthermore, the later development of body plans and organ systems in the five vertebrate classes have many clear similarities. In the second place, the mouse is a mammal, and as such is the most convenient model system for human development. Given this fact, it has special interest for us. Indeed, a number of human genetic disorders have their analogues in identified mouse genetic syndromes. The recently devised technique for genetic transformation of the mouse germ line may prove applicable to humans eventually and increases the relevance of mouse biology to human biology.

For mouse genetics, the primary genetic source material consists of nearly 300 highly inbred lines. These strains contain numerous genetic factors of interest. By making the appropriate crosses between them, it is often possible to identify the number and genetic map positions of the genes responsible for particular developmental conditions. To date, over 400 genetic loci in the laboratory mouse have been identified. A large number of these markers are biochemical, the structural genes for particular proteins. The use of such biochemical markers in developmental studies often allows genetic, molecular, and cytological analyses to be combined. An additional feature of the mouse that provides relatively easy experimentation is its relatively short generation time of 90 days.

In this chapter, we will look at the first stages of mouse development, from the beginning of oocyte development through the period of cleavage and independent embryonic growth. By the end of cleavage, when the embryo consists of about 120 cells, it contains two major cell populations, the outer trophectoderm (TE) layer and an enclosed group of cells, the inner cell mass (IM. This preimplantation embryo is termed a "blastocyst." It is the blastocyst that implants in the uterine wall, beginning the phase of maternal-dependent growth and development. The embryo proper (the fetus) eventually develops from cells of the ICM; the later postimplantation stages of development will be taken up.

The primary focus in this chapter will be on the nature of the morphogenetic system in the early embryo and its origins in the oocyte. The mouse oocyte has a special interest in this respect. In recent times, the mouse embryo has been viewed as a completely plastic system in which all early cellular commitments are impressed on essentially identical blastomeres by environmental forces. In effect, the oocyte has been interpreted as having no built-in morphogepetic instructions (Davidson, 1976). If this interpretation is correct and extends to the oocytes of other mammals, it would represent a major difference in the early developmental biology of mammals from that of possibly all other animal systems.

LIFE CYCLE AND EARLY DEVELOPMENT

Although mouse developmental biology is considerably more

complex than that of either the nematode or the fruit fly, its life cycle is simpler, being the place in one continuous phase, from the moment of fertilization until birth, the entire sequence taking place inside the mother's body.

Embryonic and fetal growth in the mouse takes approximately 20 days from fertilization to birth. The first 4.5 days are spent as an unattached, independent embryo, undergoing cleavage. When the embryo reaches the late blastocyst phase, it implants in the uterine wall, and the major phase of embryonic development begins. In contrast to most animals, the pace of the first cleavage divisions is very slow. However, the rates of cell division and developmental change increase following implantation, and accelerate further with the establishment of the placental connections to the mother's circulatory system and nutrient supply. During the first stages of postimplantation development, the three germ layers from within that portion of the embryonic structure that gives rise to the fetus, and by days 8-9, organ formation has begun. By 7.5 days, only 3 days after the 120-cell stage, the embryo proper consists of approximately 10^5 cells.

Because the embryo develops inside and attached to its mother's body, a special issue arises in mammalian development; the role of the mother in the developmental process. Is there a continuing "instructive" maternal role *in* development? This question has been investigated by culturing embryos in various nonuterine (ectopic) sites, particularly in males. Although such embryos show some abnormalities, the results indicate that the embryonic developmental sequence is governed by the embryo itself; the sole function of the maternal environment is to provide adequate nutrition to the developing individual. Nevertheless, the question of maternal environment is pertinent in assessing putative maternal effect mutants, since early maternal effects on oocyte development must be distinguished from later maternal influences on embryonic development. This distinction can be made by transferring early embryos to surrogate mothers of a different strain, or, if preimplantation embryos are being tested for such effects, by comparison with control embryos in vitro culture.

In general, preimplantation development is most readily analyzed in vitro. To obtain sufficient numbers of embryos for study, females are induced to "superovulate." This action is produced by injection with gonadotrophic hormones prior to mating. Normally, a female will release up to 10-20 eggs per ovulatory cycle; superovulated females can release several times this number of mature ova. The fertilized eggs can be readily removed from the female reproductive tract for further experimental analysis.

GENETICS AND GENOME STRUCTURE

The diploid chromosome number of Mus Musculus is 40. Of the 20 chromosome pairs, 19 are autosomal and one is the sex chromosomes. Sex determination in the mouse is by an X-Y chromosome system, with XX individuals being. female and XYs male, as in Drosophila. However, a fundamental difference in sex determination exists between the fruit fly and mammals. In the latter, sex is determined by the presence or absence of the Y chromosome, which establishes maleness, rather than by the Z: A ratio, as in Drosophila.

The genetic nomenclature is similar to that of Drosophila, Genes are given italicized symbols that are usually abbreviations of their names. In general gene designations begin with a capital letter unless the gene was first identified by a recessive mutation (e.g., qk for quaking).

The development of chromosome banding techniques. Gbanding and Q-banding, which distinguish specific homologous chromosome' pairs, made possible the assignment of genes to particular chromosomes in the karyotype. This method uses translocations that move known genetic marker from one chromosome to another, in combination with an analysis of the changes in karyotype structure produced by the translocations, as detected by the banding techniques. (Until the advent of the chromosome banding techniques, genes could be assigned only to genetic linkage groups, but the identity of particular linkage groups with particular chromosomes could not be ascertained.)

A map of the mouse genome is given in the Appendix. All chromosomes are acrocentric and numbered from largest to smallest. The sex chromosomes are simply designated X and Y. The Y chromosome is similar in size to the smallest autosome pair (chromosome 19) but appears to be empty of genes, aside from male fertility genes and the locus for the H-Y antigen (a surface male-specific histocompatibility antigen). The nomenclature for chromosomal rearrangements is similar to that of *Drosophila.*

The molecular organization of the mouse genome raises many of the same fundamental questions encountered in connection with the genomes of the nematode and the fruit fly. The major gross difference between these genomes is in size. At 3×10^9 base pairs, the haploid mouse genome is nearly 20 times larger than *Drosophila.* Of this total 8-10% consists of highly repeated satellite DNA, principally a 140-bp, adeninethymine (AT)-rich repeat, reiterated 10^6 times in the centromeric heterochromatin. The single copy fraction is approximately 60-76% and the midrepetitive fraction is 15-25% of the genome.

The relative dispositions of single copy and repetitive sequences in the mouse genome are the subject of some dispute. The standard interpretation of eukaryotic interspersion patterns has been that there are two major types. In this interpretation, the most common pattern consists of single copy DNA sequences in 1000-to 300-bp stretches, interspersed with short midrepetitive sequences about 300 by in length. The alternative interspersion pattern is designated the *"Drosophila* pattern," because it was first found in *D. melanogaster.* In this arrangement, long single copy stretches (averaging more than 13,000 bp) are interspersed with long blocks of midrepetitive DNA (5000 by or more in length).

In the original studies, all mammalian genomes were judged to follow the standard, short interspersion pattern. However, *Moyzis et al., (1981a)* have reevaluated these findings and concluded that much of the apparent difference between genomes in the interspersion pattern disappears when they are compared by identical renaturation methods. As these authors stress, the categories of "single copy" and "mid repetitive" are operational groupings, and the particular renaturation conditions employed can shift sequences from one category to the other. They show that under stringent reannealing conditions, the rodent genome shows the *Drosophila* pattern, with long single copy stretches (7200 ± 2000 bp) interspersed with long blocks of midrepetitive sequences. These mid repetitive blocks, in turn, consist predominantly of "scrambled clusters" of shorter (300-bp) blocks. Interestingly, an analysis of the long blocks of midrepetitive sequences in *Drosophila* shows that many of these too consist of similar scrambled clusters.

The functional significance of the sequence interspersion pattern remains unknown. For the reasons discussed in convection with the structure of the *Drosophila* genome, individual midrepetitive sequences are unlikely to have major, direct roles in controlling the expression of individual genes. However, it is not impossible that they have some general function in chromosome folding pattern or replication that indirectly influences gene expression. Conceivably, germ line transformation with engineered genes altered in neighbouring midrepetitive sequences, followed by analysis of their expression patterns may help to answer these questions.

OOGENESIS, FERTILIZATION AND PREIMPLANTATION DEVELOPMENT

Like all animal development, the beginnings of mouse development are to be sought in the structure and properties of its oocyte. The development of the mouse oocyte takes place in two discrete phases. The first phase extends from midembryogenesis until 5 days after the birth of the female mouse. It lasts from the formation of oogonial stem cells by the primordial germ cells to the formation of immature oocytes, which are held in late diplotene of meiosis I. The second phase begins with the attainment of sexual maturity and involves oocyte growth and maturation. This takes place in small batches of oocytes within each estrous cycle.

In its general aspect, the initial development of the ovary in the mouse bears a certain resemblance to that in *Drosophila*. The primordial germ cells, which look identical in the two sexes, are distinguishable from other cells by their round shape and diffuse chromatin, characteristics similar to those of pole cells in *Drosophila;* however, they are most readily detected and tracked by their high alkaline phosphatase content. As in the fruit fly embryo, these primordial germ cells undergo a major migration from their posterior site of origin to the gonad rudiment. This migration, between 8 'days postconceptus (pc) and 13.5 days pc. takes place from the caudal end of the embryo proper (the primitive streak) through the hindgut to the genital ridges (the somatic rudiment of the gonad). In female embryos, the arrival of these primordial germ cells stimulates the proliferation of the epithelial cells of the genital ridges; the primordial germ cells themselves also commence multiple rounds of cell division, to form oogonia. The oogonia subsequently undergo two waves of inward migration. Most of those in the first wave degenerate, but the cells of the second wave come to rest in the peripheral (cortical region of the developing ovary. These cortical oogonia continue to proliferate, although remaining connected by many intercellular bridges. As we have seen, this syncytial arrangement is common in early germ line cells, and presumably serves to facilitate intercellular molecular transfer and synchrony of nuclear divisions.

During the migration to the ovarian cortex, groups of oogonia become surrounded by prefollicle cells of mesodermal origin. Eventually, each pro-oocyte is covered by a monolayer of these somatic cells to form a "unilaminar follicle.." Prooocytes enter the first meiotic division but arrest at late diplotene until just prior to ovulation. Development of the unilaminar follicle is accompanied by the loss of syncytial connections between pro-oocytes. The proliferation of follicle cells and the secretion of extracellular material serve to isolate the follicles further from one another.

Oocytes remain quiescent until sexual maturation, when they are hormonally stimulated to commence growth and development in small groups. Although the estrous cycle takes 5 days, each preovulatory growth periodlasts for 20 days, with most growth taking place between 20 and 6 days prior to ovlulation. During this period, the growing oocyte increases in diameter from 20 to 80 gm and the surrounding follicle cells multiply to reach 50,000 per follicle. During these final stages, these final stages, the oocyte comes to occupy fluidfilled chamber, the "antrum," within the follicle. The follicle cells act as feeder cells for the oocytes; 85%

or more of the metabolites required by the oocyte are passed to it through the follicle cells.

The postnatal phase of oocyte development involves both growth and differentiation of the cell. A characteristic sequence of differentiative changes in the major cytoplasmic organelles-ribosomes, endoplasmic reticulum, nucleoli, mitochondriaoccurs, and marked protein synthesis takes place in both oocyte and helper

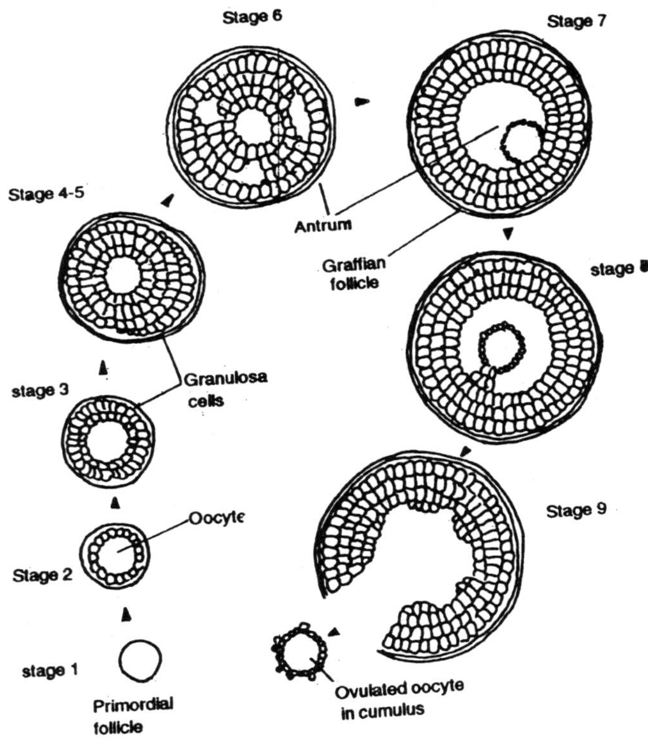

Figure 9.1 : Stages of mouse oocyte and follicle growth.

cells. As in many other animal oocytes, synthesis and accumulation of histone proteins and of tubulins occurs. These proteins are subsequently utilized in cleavage. The tubulin subunits may comprise as much as 1% of the total protein in the mature oocyte, thus constituting one of the most abundant protein fractions in the egg (as they are in the fruit-fly oocyte).In common with other cellular differentiations, the oocyte growth phase features stage-specific alterations in the pattern of polypeptide synthesis, as assayed in two-dimensional separations.

This period is also one of active transcription, of both ribosomal and informational RNAs. The synthesis of poly A^+ mRNA is part of this transcriptional program. When [^{3}H]uridine is used as the radioactive label, 9.6% of the total counts in the mature oocyte are found in poly A^+ mRNA. Indeed, the fraction f stored RNA that is in the form of mRNA is substantially higher in the mouse ovum than in the sea urchin or in the frog *Xenopus,* where the proportion is closer to 1-2% of total stored RNA.

The period of oocyte growth and differentiation ends about *6* days prior to ovulation and is succeeded by a 5-day phase of follicle cell proliferation and antrum development. In the last 9-10 hr before ovulation, the oocyte reenters meiosis and completes the first meiotic division, giving rise to the first polar body. The oocyte nucleus then proceeds directly to meiosis II without re-formation of the nuclear membrane but halts at metaphase until fertilization. Ovulation involves release of the ovum from the follicle into the oviduct. It is surrounded by a polysaccharide-rich envelope, termed the "zonapellucida," and a group of follicle cells, the "cumulus oophorus." Because receptivity to mating closely correlates with ovulation, the time of fertilization can be gauged fairly accurately under conditions of controlled mating. Fertilization involves penetration by the sperm through the cumulur oophorus, which is shed during passage down the oviduct, and the zona pellucida. Following entry, the sperm head is converted to the male pronucleus and loses its complement of chromosome-coating protamines, which are replaced by histones and acidic proteins from the maternal cytoplasm. Fertilization triggers resumption of meiotic progression in the oocyte nucleus; completion of meiosis is quickly followed by extrusion of the second polar body. The two pronuclei then move slowly into apposition. (DNA replication takes place in the pronuclei and is completed during this lengthy migratory phase.).

During the first 3 days following fertilization, the zygote progresses through four cleavage divisions to the 16-cell stage, the morula. In the following day and a half, the solid ball-like morula develops into the blastocyst. The blastocyst is a hollow, nearly spherical fluid-filled ball of cells, and, as mentioned earlier, it is both the implanting structure and the carrier of the precursor cells of the true embryo.

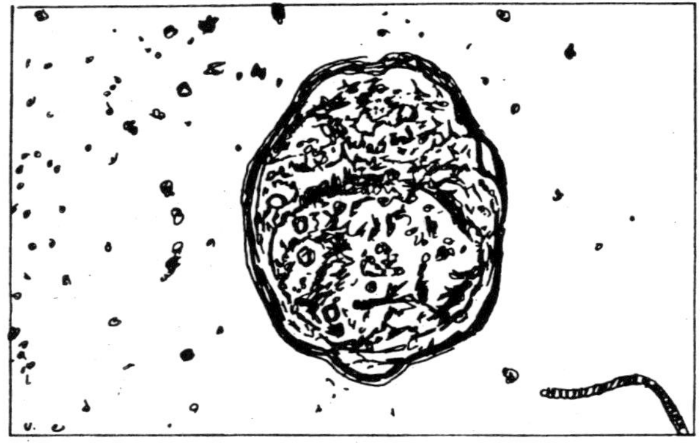

Figure 9.2 : An expanded 3.5-days blastocyst. diameter 70-80 μm.

Cleavage in mammals is a slow process. In the mouse, the first cleavage does not take place until 24 hr postfertilization, after the prolonged migration and fusion of the pronuclei. Successive divisions take place at approximately 12-hr intervals. The eight-cell stage may be a developmentally critical juncture. It is at this point that the heretofore loosely knit blastomeres undergo the process of compaction. During compaction, the cells transform from spherical to columnar and develop a variety of intercellular junctions. The net result is the drawing together of the ball cells. Compaction is-almost certainly important in creating certain properties of polarity within the blastomeres of the eight-cell state; the significance of such polarity will be discussed later.

From the eight-cell stage on, cleavage results in the formation of distinct inner and outer cell populations. By the 16-to 32cell stage, the inner cells have begun to secrete fluid into the developing central cavity, the blastocoele,. By the 60-cell stage, approximately 3.5 days after fertilization, the embryo is at the early blastocyst stage and consists of two cell populations. The outer layer of approximately 45 cells constitutes the trophectodermal (TE)) layer and the internal 15 cells comprise the ICM. Formation of the early blastocyst is followed by reexpansion of the embryo to fill the zona pellucida.

The final phase of preimplantation development involves the production of two further cells groups. The ICM gives rise to group of endodermal cells that eventually cover the entire inside of the blastocoele. And the TE diverges into a cap of "polar TE" cells, which comprise the sides of the blastocyst. By the late blastocyst stage, some of these mural TE cells have begun to become polyploid; eventually, these large polyploid cells become the "primary giant cells" of the implanting blastocyst wall.

EXPRESSION OF THE ZYGOTIC GENOME DURING PRE IMPLANTATION DEVELOPMENT

The changeover from maternal to zygotic genome control in the early mouse embryo has been tracked by several means. One of these is analysis of the stores of maternal mRNA and their depletion during preimplantation development. As noted above, the poly A mRNA pool is unusually large, amounting to approximately 10% of the total RNA in the zygote. This maternal RNA pool is found to decline in parallel with the prelabeled maternal RNA. The measurements show an initial depletion of prelabeled mRNA to 60% of the prefertilization ;ontent during the first 24 hr of development (to the two-cell stage), followed by a more gradual decline between the cellcell and early blastocyst stages to a final value of .30%. The actual decline may in fact be considerably greater; inhibitor experiments and tests of in vitro translation of extracted maternal mRNA indicate a rather rapid inactivation or depletion of these stores by the mid-two-cell stage.

Despite the large starting maternal mRNA pool, the embryo becomes dependent on transcription from its own genome early in development. Blockage of transcription by either actinomycin D or a-amanitin stops development at any point after the first cleavage and reduces amino acid incorporation from the two-cell stage on. This process is significantly different from the early developmental programs of sea urchin and amphibian embryos. In these animals, actinomycin D treatment during cleavage has little immediate effect on amino acid incorporation or cell division. However, mRNA synthesis is shut off and development arrests at the postcleavage stage.

In accordance with the inhibitor study results, labeling of new RNA synthesis by radioisotope incorporation shows that zygotic genome transcription begins early. Using [^{3}H]uridine incorporation, *Knowland* and *Graham* (1972) placed the beginning of transcription in the embryo at the middle of the two-cell stage. With the same label. Levey et al. (1978) similarly located the beginning of nonribosomal nuclear RNA and of poly A+ mRNA at this point. However, when [^{3}H]adenosine, which is taken up more efficiently than uridine, is used, the synthesis of new species of nuclear RNA and poly A+ mRNA is found to begin in the middle of the one-cell stage. Nevertheless, formation of labeled poly A^+ messengers at the one cell stage involves primarily addition of poly A tails to preexisting maternal mRNAs.

Given the early expression of the zygotic genome, when does the cytoplasmic composition of the embryo begin to change over from maternally encoded products to those characteristic of the embryonic genome? Bio-chemical experiments suggest that the transition begins no later than the two-cell stage. However, results of both inhibitor and labeling experiments are subject to various interpretations. Independent estimates for ascertaining when the zygotic genome begins to affect embryo cytoplasmic composition can be obtained by genetic means. Such tests involve detection and discrimination of paternal genome-encoded products from maternal ones, while maternal products may derive either from the oocyte or from the maternally derived chromosomes of the zygote, paternally encoded products in the embryo can come only from the zygotic genome.

In *Caenorhabditis* and *Drosophila,* it is possible to test for paternal gene activity by determining whether rescue from lathal maternal effects takes place in embryos with wild-type paternal alleles. In the mouse, where few maternal effect mutants are known (see the following section), one must resort to biochemical methods for detecting such paternal gene expression. One such method is to score for a quantitative increase in gene expression programmed by a paternal gene. An example is provided by a test for B-galactosidase expression in early embryos, an enzyme encoded by a gene — on chromosome 9. During development to the blastocyst stage, this enzyme, like that of two other lysosomal hydrolases, undergoes a 50-to 100-fold increase in activity. This activity increase in mediated by a closely linked, cis-dominant

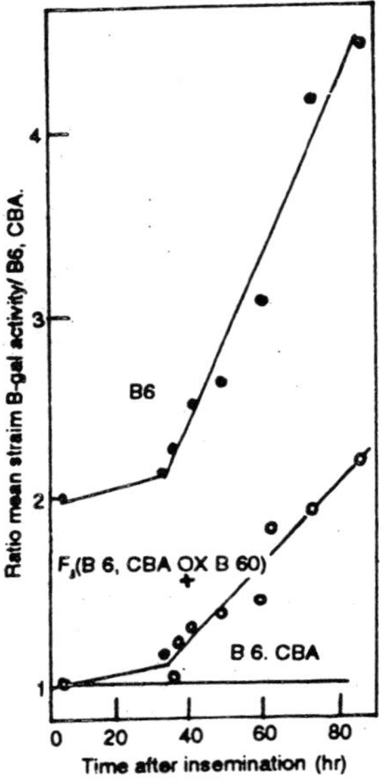

Figure 9.3 : Time course of paternal expression of β-galactosidase in mouse embryonic development.

regulator gene site, designated *Bgl-s.* One allele of this cis-dominant regulator, *Bgl-s,* produces a twofold higher level of activity than the other known allele, *Bgl-s. To* establish the time of paternal gene expression for B-galactosidase in embryonic development, *Esworthy* and *Chapman* (1981) mated female mice homozygous for the *Bgl-e* (low activity) allele to males with the *Bgl-s'* (high activity) allele and collected embryos at successive times after fertilization for measurement of enzyme activity. The strains employed were nearly identical in genotype ("congenic"), except for the region on chromosome 9 containing the structural gene and its linked regulator.

Figure shows the plotted increase in enzyme activity in hybrids as a function of developmental stage when normalized to the activity increases seen in parallel Bgl-sd x Bgl-sd crosses. The result show a paternal allele-mediated increase in activity that extrapolates back to 36-45 hr, or shortly after the second cleavage division. That the increase mirrors a true turn-on of gene expression rather than an effect of heterosis is shown by the reciprocal cross (with a high maternal background) results, which did not give a comparable increase. The findings indicate that the cytoplasmic composition of the embryo begins to reflect zygotic genome expression by the two-cell stage. Comparable tests with the chromosome 5 structural gene for B-glu-curonidase give a similar result.

From the two-cell stage on, zygotic genome expression accelerates, its products increasingly dominating the cytoplasm Ribosomal RNA synthesis cannot be detected in the one-cell embryo, but by the mid-two-cell stage it accounts for a significant fraction of the new transcripts. General transcription rapidly accelerates in the eight-cell embryo, the point at which compaction occurs. By the early morula state, rRNA synthesis is approximately 16-fold more rapid than in the two-cell stage.

In contrast, protein synthesis is active throughout early development and even in the unfertilized oocyte. Much of this early synthesis is carried out on maternal templates. However, detailed analysis, employing two-dimensional gel electrophoresis, shows that zygotic genome-encoded protein products are detectable by the early two cell stage.

One of the most interesting findings to come from such protein synthesis studies concerns the occurrence of postranscriptional regulation in the maternal mRNA pool These regulated messengers code for a group of proteins of 35 kd, which first appear during the late one-cell stage. The scheduled synthesis of these proteins continues unabated in the presence of the transcription inhibitor a-amanitin, at a drug concentration sufficient to inhibit all in vivo RNA ploymerase II activity. Furthermore, when maternal mRNAs are extracted from unfertilized eggs and placed in an in vitro protein-synthesizing system, these proteins are made in quantities comparable to those of two-cell embryos. If follows that the responsible mRNAs must be held in some state that is unfavourable for translation, until the early two-cell stage, and are then released by some signal to the translation apparatus. One possibility is that this signal is a tinted polyadenylation of these messengers, permitting transport out of the nucleus, because the onset of translation correlates with the approximate period in which long poly. A tails are added to preexisting maternal mRNAs.

The cellular functions of these posttranscriptionally controlled proteins and the significance of their controlled translation are unknown. We noted earlier the existence of a parallel phenomenon in *Drosophila.* A third example of such posttranscriptional control has been reported in the sea urchin, involving delayed translation of a histone H3 maternal mRNA. One possible explanation for the phenomenon is that such posttranscriptionally regulated mRNA s might code for proteins required in abundance by the early embryo, and therefore must be supplied by maternal templates, but that their "premature" translation could cause some kind of damage to the embryo. This suggestion might be tested by examination of the effects on development of early injection of the purified proteins.

The question of mechanism is also unresolved but may involve different processes in different organisms or even for different mRNA species within the same animal embryo. *Drosophila* involves sequestration in cytoplasmic RNP particles, but the sea urchin histone mRNA seems to be retained by some selective mechanism in the nucleus. Delayed polyadenylation might, in principle, be a general mechanism of posttranscriptional control in the mouse, but the particular MRNAs that undergo delayed translation are stored as polyadenylated species.

Although changes in electrophoretic patterns of labeled polypeptides can be used as an index of changes in gene expression, they are not an infallible guide to changes in protein synthesis. About 50% of the changes occurring in early preimplantation development, as registered on two-dimensional separations, result from reproducible patterns of posttranslational modification—the addition of nonprotein moieties, such as glucosamine and phosphate, to preexisting polypeptides that cause a change in migration behaviour. Some of these posttraslational modifications are set in motion by oogenesis itself and occur in unfertilized eggs. From the eight-cell stage, there is relatively less change in the overall distribution pattern of labeled polypeptides in two-dimensional gels from whole embryos. Thus, during the period of blastocyst formation, the overall protein synthetic pattern appears relatively stable.

Despite this relative constancy in protein synthesis between the eight-cell/compaction stage and the early blastocyst, a few significant changes in gene expression take place. The first is a key regulatory event in the late morula state, as the embryo goes from a 16-cell structure to one of 32 cells, which is necessary for the secretion of fluid that takes place during blastocyst formation. This fluid initially accumulates between cells and then collects in the developing blastocoele, the entire process being referred to as "cavitation."

Smith and *McLaren* (1987) have shown that the onset of cavitation requires five DNA replication cycles. Presumably, the signal to begin cavitation involves either a direct "counting" of these replication cycles or the attainment of a characteristic nucleocytoplasmic ratio during cleavage. In investigating the molecular biology of this developmental event. *Braude* (1989) found that the onset of cavitation requires-transcription and concluded that this transcriptional event occurs during or as a result of the fifth replication cycle. The nature of the lint. between the fifth replication cycle and transcription has not been elucidated. Since zygotic genome transcription has commenced well before this point, the control event cannot be a simple titration of transcription inhibitor by DNA as occurs in *Xenopus*.

The second set of events concerns the divergence of the first two distinct cell populations, the ICM and the TE cells. *Van Blerkom et al.* (1986) were the first to detect differences in patterns of polypeptide labeling between these cell groups. They dissected ICM and TE cells from 3.5-day (expanded) blastocysts, incubated the two fractions separately with [^{35}S]methionine, and then analyzed the patterns of labeled polypeptides with two-dimensional separations. Although the majority of spots were shared between the two cell groups, as expected from the general constancy of labeled polypeptides in whole embryos between the 8-and 64-cell stages, a few distinctive ICM and TE spots were detected. Thus, by the late blastocyst stage, the two cell populations, which are recognizably different in position and beginning to be distinct cytologically, can be distinguished by molecular markers.

Handyside and *Johnson* (1988) subsequently showed that the "ICM-Specific" and "TE-specific" cells are first detectable between the early morula (12-to 25-cell) and late morula 25-to 30-cell) stages. When the inner cells from embryos of these stages were isolated by the technique of immuno-surgery these inside cells were found to label just the shared and ICMspecific polypeptides and not the TE-specific ones. Some of these differences may reflect posttranslational modification rather than new gene expression per se, but the significant point is that the molecular events of differentiation are detectable before there are cytologically distinct ICM and TE cell populations.

The sequence of changes in gene expression and macromolecule composition that take place in the preimplantation embryo. The entire set of changes constitutes a complex picture. However, the salient feature is that zygotic genome expression begins early in this embryo, and zygotic genome products increasingly dominate the molecular composition of the embryo from the two-cell stage on.

DETERMINATION EVENTS IN 1 HL PREIMPLANRATION EMBRYO: GENETIC ANALYSIS

Investigation of determination mechanisms in the early mouse embryo has relied principally on the first method, chimera construction. Extensive mutant hunts are impossible because of the limit number of progeny that can be screened; furthermore, the collection of existing mutants includes extremely few with maternal effects. However, the chimera experiments alone have proven instructive about the timing of determination events in the preimplantation embryo. This information, in turn, provides some clues to the mechanisms of determination in this embryo. We will first look at the methods of chimera analysis used in the mouse and the application of these techniques to analysis of early determination. In the next section, we will examine some of the hypotheses about determination in the mouse embryo that have been proposed.

In *Drosophila,* the analysis of early embryonic determination has involved two principal kinds of genetic technique. The first is the creation of genetic chimeras by removal of genetically marked cells from the blastoderm stage and subsequent culture in recipient embryos or larvae. The fates of the donor cells in these chimeras are used to establish the timing of the first determinative events. The second genetic technique is the isolation of maternal effect mutants that produce aberrations in the morphogenetic system of the egg. These experiments have permitted a genetic characterization of the mechanisms of determination.

Chimera Construction and Analysis; Methods

There are several methods of constructing chimeric mouse embryos. The first early embryo aggregation was invented by *Tarkowski* (1991) and *Mintz* (1992). In this technique, eight-cell embryos from two genetically distinguishable strains are stripped of their zonae (mechanically, or now usually by digestion with pronase) and then brought into contact; fusion normally occurs within a period of 2-4 hr at 37°C. The composite morulae are then transferred to the uterus of a pseudo-pregnant foster mother, where implantation and subsequent development take place. (Such foster mothers are typically prepared by prior mating to vasectomized mice.) A substantial proportion of such morulae develop and give rise to normal newborn mice. In experiments in which the two input genotypes differ in their coat color phenotype, the coat color pattern of the chimeras can have a striking appearance. Note: the term "allophenic mice" is also found in the literature to denote chimeric mice.)

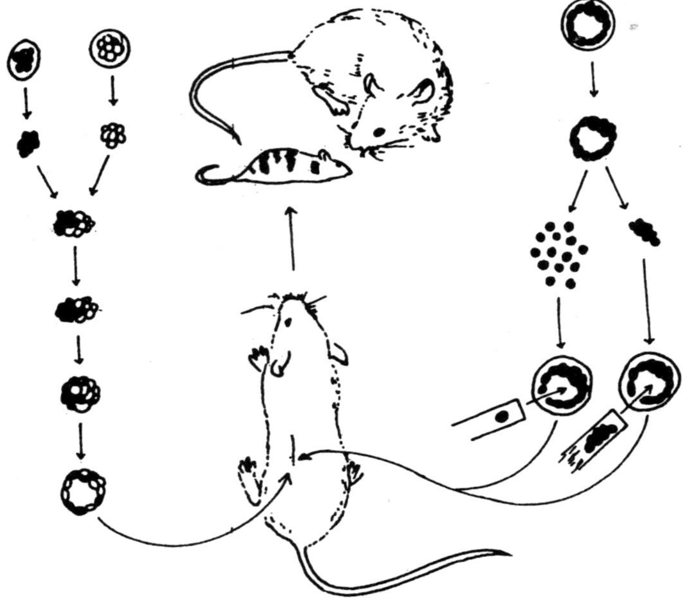

Figure 9.4 : Principal methods of chimera production in the mouse. Morula aggregation shown on the left, blastocyst injection on the right.

The second method, introduced by *Gardner* (1988), permits the assay of the developmental capabilities of cells of different stages, either singly or in small groups, without the major disruption of preimplantation development that embryo aggregation entails. In this procedure, single or multiple cells of the desired stage and genotype are injected into host blastocysts through a narrow incision. The recipient blastocyst is then placed in a pseudopregnant foster mother; those that have not been irreparably damaged by the injection procedure give rise to chimeric fetuses. The method allows the testing of developmental capabilities of any of the cells in the preimplantation embryo. The method can also be used to look at developmental potentialities of cells in postimplantation stages.

A potential drawback of the cell injection method is that the blastocoels environment may affect the developmental paths taken by the injected cells. To avoid this complication, a third method has been invented. In this scheme, the cells to be tested are aggregated with eight-cell morulae to create chimeric embryos.

The informativeness of all chimera experiments depends on the sensitivity or resolution with which the cells of different genotypes can be resolved. When chimeric embryos are allowed to develop to birth, the use of coat color markers is adequate, at least for detecting respective contributions to the epidermal tissues. However, when internal tissues are to be traced, or when intermediate stages of development are to be examined, histochemically distinguishable genetic markers must be used. These are usually enzymatic or antigenic markers.

Biochemical markers are broadly classified as either *direct* or *indirect*. With direct cell markers, individual cells can be scored in tissue sectibns, permitting fine-scale tracing in situ of the location and relative contribution of the two genotypes. Plus or minus activity enzyme activity differences and surface antigenic markers are in this category. Indirect cell markers are those that can be scored only in bulk tissue preparations. The most usual indirect genetic marker is an electrophoretic difference in an enzyme, that is electromorphic alleles. With indirect cell markers, the chimeric embryos are dissected into their component tissues, which are ten assayed separately for the two genotypic contributions. The two most commonly used indirect markers are electromorphic differences for the enzymes isocitrate dehydrogenase (IDH) and glucose phosphate isomerase (GPI). For the latter, minor electrophoretic differences can be detected in proportions as small as 1-3% of total activity.

In general, direct markers give more information with less chance of error. With indirect markers, there is always the problem of cross-contamination of tissue samples by cells of the other genotype. However, it is not always possible to find strains differing in alleles for a good direct marker. The recent development of techniques for discriminating cells of differing genotype at the major histocompatibility locus (H2) of the mouse may solve this problem. Because of the lack of suitable direct markers, most experiments have relied on an indirect cell marker, principally the GPI electromorphic difference.

Chimeras and the Analysis of Early Determination Events

The analysis of determination events in preimplantation development by chimera construction is similar to that used in the analysis of *Drosophila* determination in the blastoderm. One takes cells of genotype A, from a defined stage or location, and places them in host embryos of genotype B. Subsequent development is followed, and the relative contributions of A and B to the various tissues of the later embryo nor adult are scored. If all tissue types have some donor cell (genotype A) contribution, then these donor cells must have been totipotent.. If, on the contrary, they are found in only certain tissues, the original donor cells are presumed to have experienced some previous determinative events.

The chimera technique has been used to explore three fundamental questions about determination in the preimplantation embryo : (1) does the divergence of ICM from TE entail a restriction in developmental potency of either cell type? (2) If such restrictions do occur, when do they take place? (3) Does the formation of endoderm from ICM or the divergence of polar and mural TE involve determinative events?

The topic that has received most attention is the nature of the divergence between ICM and TE cells. It has long been known that these two cell populations differ distinctly in their embryological fates. The ICM of the late blastocyst contributes directly to the fetus and to certain of the extra-embryonic membranes, while the TE contributes directly to tissues involved solely in the implantation process. By 3.5 days pc, when the blastocyst consists of 60-64 cells, these two cell populations differ in distinctive cytological features. These differences include the kinds of cellular junctions each displays, the presence of microvilli on ICM and their absence from TE, and recognizable differences in mitochondrial structure. Given the difference in fates between ICM and TE, is it based on an irreversible difference in developmental capability, and if so, when does this difference become established?

The experiments of *Rossant* (1985a) demonstrate that the cells of the ICM are already determined by the 3.5-day, expanded blastocyst stage. In these experiments, the donor ICMs were aggregated with recipient eight-cell morulae and transplanted to the uterine horns of pseudopregnant foster mothers, then allowed to develop for periods ranging from 8.5 to 18.5 days. The genetic marker used for tracing the ICM-derived cells was GPI, with the donor cells specifying the b electromorph and the recipient embryos specifying the a form. Following termination of development, the fetuses and associated embryonic structures were isolated and dissected into the ICM-derived embryo plus membranes fraction and the TE derived fraction.

The analysis of the chimeric embryos showed a substantial and highly preferential contribution of donor GPI enzyme to the ICM-derived tissues, with little or no contribution to TEderived fraction. Of the 15 embryos showing a donor cell contribution, 12 showed an exclusive contribution to the embryo plus membranes fraction. The three apparent exceptions showed only a minor contribution, which probably reflected contamination with ICM-derived tissues. The results provide firm evidence that ICM cells from 3.5-day expanded blastocysts are already restricted in their development capacity. A similar result has been reported using a direct cell marker (an antigenic difference) in rat-mouse chimeras, in which rat ICM cells were injected into mouse blastocysts.

TE cells from 3.5-day blastocysts are comparably restricted in their developmental potential. When TE vesicles are emptied of their ICM cells before fostering, they give rise to trophoblast cells (the TE cells that mediate implantation) and undergo implantation. However, they cannot give rise to ICM or to those structures normally derived from ICM. Furthermore, injection of TE cells into 3.5-day blastocysts never yields embryos with a donor cell type contribution. The results show that by the 60-to 64-cell stage, the cells of the ICM and TE are committed to distinct and restricted pathways of development.

When does this restriction in developmental potential arise? The experiment of *Kelly* (1985) provides a lower bond. She dissociated four-cell embryos in vitro and then allowed each blastomere to divide once, to give four "octet pairs. "The members of each pair were then separately aggregated with individual eight-cell carrier

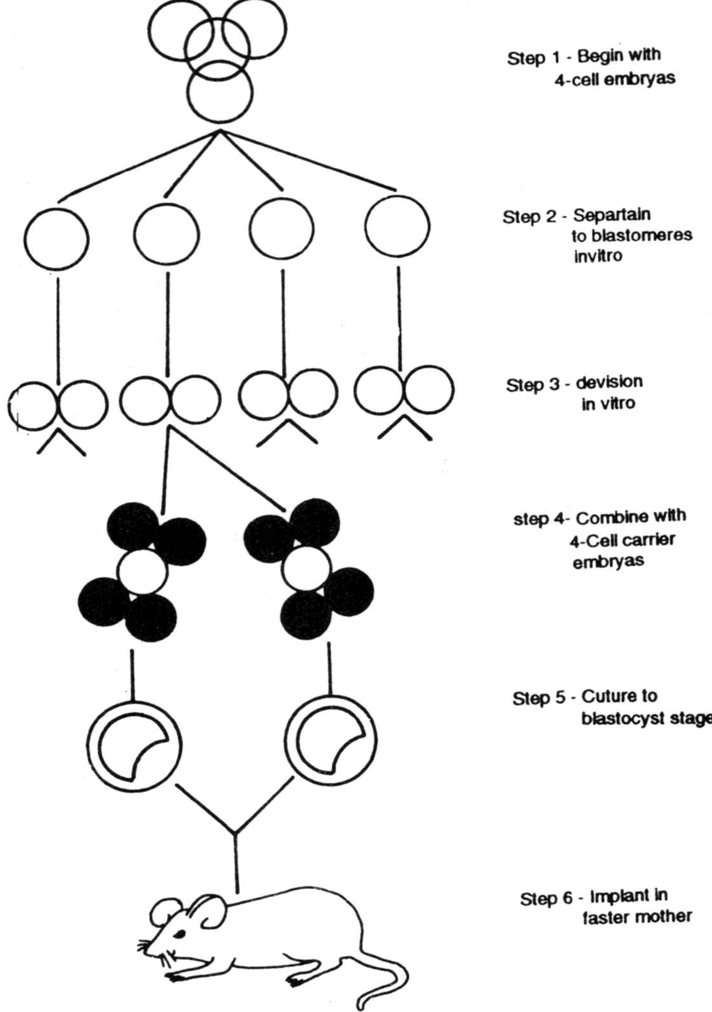

Figure 9.5 : Test for totipotency of blastomeres at the eight-cell stage in the mouse embryo. See text for details.

morulae and fostered. Donor cells carried the *Gpi-1* allelic form, while host morulae carried *Gpi*1. In addition, the donor strain was albino, so that chimeras that developed to term could be distinguished by coat color. Fourteen chimeras, derived from seven octet pairs, had donor cell contributions in the fetuses. In five of the pairs, donor cells had participated in the formation of both ICM-TE derived fractions. A substantial donor cell contribution was also seen in all chimeras allowed to develop to term. Evidently, the blastomeres of the eight-cell embryo do not show segregation of developmental potential for either ICM or TE. This work demonstrates that blastomeres at the eight-cell stage are totipotent. Therefore, the first determinative events occur between the 8- and 64-cell stages.

To pinpoint the time of developmental restriction more precisely, the technique of *"immunosurgery"* was used by *Handyside* (1988) to isolate and the test the capabilities of inner cells at successive intermediate stages between compaction and the expanded blastocyst. Immunosurgery on mouse embryos involves the selective lysis of outside cells by exposure of the embryos to rabbit anti-mouse serum, followed by washing and treatment with complement (a substance that lyses antibodyreacted cells). When the treatment is applied to embryos of successively more advanced stages and the isolated inner cells are tested in vitro for their differentiative capabilities, the inner cells of later morulae and early cavitating blastocysts are found capable of generating TE derivatives. However, the ICMs of 3.5-day expanded blastocysts do not possess this capacity and produce only endodermal layers.

A somewhat different result, obtained with the same procedure but with embryos from a different strain, was obtained by *Hogan* and *Tilley* (1988), who found some 3.5-4.0 ICMs capable of generating TE-derived structures. These authors suggested that the timing of the restriction event differs somewhat between different strains and that it is a progressive, gradual process, with some potential for reversal up to the 4.0day stage. Such gradual loss of developmental potential could reflect either a decreasing plasticity of all cells or a progressive shift in the composition of the inner cell population from those with TE developmental capacity to those that do not have it.

Do these restrictions in cellular developmental capacity have nuclear state correlates? Two sets of published results suggest that they do. *Illmensee* and *Hoppe* (1981) compared the abilities of single nuclei from ICM and TE cells of expanded

blastocysts to support development in eggs when substituted for the zygotic nuclear inheritance. They reported that ICM nuclei had a significantly higher capacity than TE nuclei to support development, even obtaining three mice that developed to term with substituted ICM nuclei. *Modlinski* (1981) performed a similar comparison without removing male and female pronuclei. He compared the ability of nuclei from morulae (12- to 18-cell embryos) to that of ICM and TE nuclei to support embryonic development by transplanting nuclei directly into newly fertilized eggs without pronuclear removal. Under these conditions, the transplanted nucleus fuses with the zygote nucleus to form a tetraploid hybrid nucleus. When nuclei from either morulae or ICM cells are tested in this manner, approximately half of the surviving transplants develop to the blastocyst stage; in contrast, TE nuclear transplants invariably stop at earlier stages. The results of these nuclear addition experiments are thus similar to those obtained with nuclear substitution.

Less is known about the divergence of cell types within the ICM and TE cell populations in the final 24 hr of preimplantation development than about the initial separation of ICM from TE. However, the delamination of the internal layer of endoderm from the ICM probably-also involves a determinative. restriction. The endoderm cells of the late blastocyst differ from the ectodermal cells of the remaining ICM in three respects: they show more intense histochemical staining, they possess a more extensive endoplasmic reticulum, and their surface membranes have a "rough" character that contrasts to the "smoother" surface of the ectodermal ICM on dissociation in vitro under the appropriate conditions.

Gardner and *Rossant* (1979) purified rough (endodermal) and smooth (ectodermal) inner cells and injected them singly or in small groups into 3.5-day blastocysts, the latter marked with a different GPI electromorph. Early endodermal cells contribute solely to the endodermal layer of the yolk sac, while ectodermal cells contribute solely to the mesodermal layer of the yolk sac and the fetus proper, but never to the endoderm of the yolk sac. (The structure of the early postimplantation embryo will be described). Evidently, by the time the two ICMderived cell populations of the late blastocyst have separated, they have diverged both in fate and in their respective developmental potency.

The case of the two TE cell populations may be subtly different. A principal difference between polar and mural TE cells is in their nuclear character Polar TE cells remain diploid and retain their division capacity, while mural TE cells become polyploid and lose this capacity. The retention of division capacity by polar TE cells seems to be a consequence of their association with the ICM in some manner, this contact serves to maintain the polar TE as diploid cells. However, while mural TE cells cannot become polar TE cells, the reverse transformation can and does occur, as polar cells are pushed to the sides and assume positions away from the ICM. Determination in the TE cell populations is this a one-way street, not the mutually exclusive pair of possibilities exhibited by the other divergences.

Mutant Analysis of Preimplantation Development

A genetic analysis of the mechanism of determination in the early embryo requires maternal effect mutants-specifically, those affected in the morphogenetic system of the egg. To date, no such mutants have been identified among the very few maternal effect mutants known in the mouse. However, among the few known and suspected maternal effect mutants of the mouse, none has yet proven informative about the morphogenetic system of the egg or about distinctive control events in the early embryo.

The dearth of mouse maternal effect mutants at first seems puzzling Maternal effect mutants are abundant classes in C. elegans and Drosophila. As we have seen, minimal estimates of the number of loci in these organisms that can mutate to this phenotype are in the range of 10-20% of the total gene set; the real proportion of gene functions expressed in oogenesis its probably considerably higher in both of these animals. Oogenesis in the mouse involves as complex a cellular differentiation as any in this animal. Even if the proportion of genes expressed in oogenesis is smaller in mammals than in invertebrate systems (because mammals have more genetic functions), the absolute number of oogenesis genes and hence potential maternal effect genes should still be large. Indeed, there are very few early developmental mutants in the mouse at all, although zygotic genome expression begins early and is essential for preimplantation development from the early cleavage divisions.

The explanation is probably connected to the particular features of gene expression in early development, and specifically to the *relative demand* for certain gene products in oogenesis early development in comparison to later stages. Many of the partial maternal effect genes in *Drosophila,* those that function in other stages of development, are first detected as maternal effect mutants precisely because the demand for their gene products is accentuated during early embryogenesis relative to later stages. In the fruit fly, cleavage and early development occur very rapidly and make large metabolic demands on the stored materials in the embryo. In the embryo of the mouse, the situation could easily be the reverse. Cleavage and early development are leisurely processes compared to subsequent development. Under these conditions, a hypomorphic mutant of a gene expressed during both early and later developmental stages might show the more severe deficiency at the later stage and might therefore be classified as a later lethal.

It should eventually be possible to estimate the number of genes that are expressed in oogenesis and early development relative to later stages by molecular methods. As more cloned sequences of genes expressed in later stages become avilable, it will be possible to test for the presence of their transcripts in the later oocyte or early embryo by the technique of in situ hybridization. Conversely, if and when cloned sequences are made by copying mRNAs of early embryos, the same procedure can be used to estimate the proportion of early expressed genes that are also expressed at later times. Such studies may help to illuminate some of the general features of the patterns of change in gene expression from early development to later stages, a prerequisite for framing a theory of gene expression control in development.

Nevertheless, genetic methods must remain an adjunct to the molecular studies for ascertaining which functions are truly essential in early development and in which cells they are required. As methods of mutagenesis in the mouse improve, more early zygotic mutants should become available; their characterization can be expected to expand greatly our understanding of early development in the mouse.

An illustration of the genetic analysis of an early zygotic mutant is provided by a chimera analysis of the *yellow* (AY) mutant, the classic mutant of mouse genetics. Early work established its time of action as within the, period of preimplantation development. The experiments also showed that the effects on blastocyst development are delayed if the homozygous mutant embryos are allowed to develop in wildtype mothers, suggesting that part of the lethal effect involves the maternal environment of the developing embryos. However, the controversy over Ay has concerned the principal site of action of the gene. The mutant allele could be a general cell lethal, or it might act initially in either the ICM or the TE, with secondary effects on the other cellular component. Direct observation of developing homozygous mutant embryos has only produced conflicting interpretations and failed to settle the issue.

The controversy is potentially resolvable through chimera construction. The principle of such experiments is the same as that of the reciprocal pole cell transplantation in *Drosophila.* If one chimeric combination produces a mutant phenotype but the reciprocal combination does not, the locus of gene action is directly identified in the phenotypically mutant chimera. In practice, the main difficulty is similar to that in the *Drosophila* pole cell transplantation-knowing which embryos from the donor strain are the homozygous mutant ones. For any recessive lethal, the homozygous mutant individuals can be obtained only by intercrossing the heterozygotes (m + X m/+) which comprise only one-quarter of the progeny.

In the particular case of yellow, the homozygous mutants are obtained by intercrosses such as AY/ae X AY/ae. where ae is the viable allele (it produces a black coat color). In 3.5-day donor embryos, before the lethal effect sets in, the three genotypes (homozygous yellow, homozygous viable, and heterozygotes) are indistinguishable.

Papaioannou and *Gardner (1989)* employed a statistical approach to avoid this problem. They used the progeny from two crosses as their sources of ICM and TE components : AY/ ae X AY/ae (the experimental cross) and AY/ae X ae/ae (backcross). The intercross produces one-quarter homozygous mutant embryos, while the backcross produces none. A difference in embryonic yields from ICM or TE components between the two crosses can signify the cell group in which the lethal mutation is creating its effect. In experiment 1, the two sets of ICMs were tested and compared by being placed into surgically emptied blastocysts of a control albino strain (CFLP). In experiment 2, the two sets of TE cells were tested by emptying the blastocysts of their ICMs and placing control (CFLP) ICMs inside the shells. The two pairs of reciprocal chimeric embryo groups were then transplanted to hormonally prepared foster mothers and analyzed for blastocyst development, measured as the ability of embryos to produce decidual swellings (implantation).

The detailed predictions for the various possible outcomes are given in Figure If AY /AY are defective in either ICM *or* TE, then the two experiments should yield nonequivalent results: the chimeras with a defective mutant component should show a 25% deficit of chimeras from the intercross relative to the backcross embryos. As an additional control, the overall rates of chimera production (using donor GPI as a marker) within the surviving conceptuses were assayed.

DETERMINATION IN THE PREIMPLANRATION EMBRYO: PROPERTIES AND MECHANISMS

A key question about the events that separate ICM from TE is whether they are initiated by some form of preexisting organization in the mouse oocyte. A specific hypothesis of this kind was initially formulated *by Dalcq* (1987) on the basis of fixed preparations of mature mouse oocytes. He reported that these specimens exhibited differentiated "dorsal" and "ventral" cytoplasm and proposed that these cytoplasms were differentially segregated to blastomeres by the eight-cell stage. In the Dalcq hypothesis, the dorsal cytoplasm eventually gave rise to the ICM and the ventral cytoplasm to the TE.

It now appears, in fact, that the regional differentiation of fixed eggs reported by Dalcq was an artifact of the fixation process. Furthermore, the idea of a strict separation of ICMand TE -forming potential .was disproved by the experiments *of Kelly* (1985), described above, which demonstrated that blastomeres at the eight-cell stage are equally capable of giving rise to ICM and TE cells. Although the observations of totipotency at the eight-cell stage do not eliminate the possibility that a segregation of ICM and TE "determinants" occurs at a later stage, this too appears unlikely. The molecular and embryological evidence suggests that the developmental restrictions that accompany separation of ICM from TE take place relatively gradually and with a certain degree of reversibility or lability.

If the determinative process is not rigidly fixed by the cytoplasm that each blastomere inherits from the egg, then the divergence of ICM from TE must involve some form of differential cell-cell or cell-environment interaction. The obvious physical differences between these two initial cell groups is in their relative positions: the ICM cells derive from cells that are inside the blastocyst, while the developing TE cells are at all times outside and exposed to the environment. As first noted by *Mintz* (1964), this difference in physical position constitutes a difference in "microenvironment" that might shape the initial divergence in fate and developmental capacity between the two cell types.

The first experimental findings to support this "insideoutside" hypothesis of determination were presented by

Tartowski and *Wroblewska* (1987). They examined the products

of in vitro development of isolated blastomeres from both four-cell and eight-cell embryos and concluded that all cells at these stages had the capacity to give rise to vesicular (blastocystlike) forms. Thus, there was no suggestion of a segregated ability to form ICM specifically from any of the baastomeres of these two stages. Significantly, however, the isolated one-and two-cell products of eight-cell embryos were found to form a higher percentage of purely trophoblastic vesicles (lacking ICM) than did blastomeres from four-cell embryos. This difference is explicable if fluid secretion from the outside cells, and hence blastocyst formation, tends to occur only after a particular number of cleavage divisions-for instance, after the fifth cleavage division. Under these circumstances, cell clusters derived from one out of eight blastomeres (i.e., single blastomeres from eight-cell embryos) would necessarily be smaller at the tirese of fluid secretion than mini-embryos derived from one out of four blastomeres and would be less likely to have inside (proto-ICM) cells. In this view, the capacity to form ICM is solely a function of being inside at the time fluid secretion begins.

The strongest experimental support for the inside-outside hypothesis comes from chimera experiments in which cells from four-or eight-cell embryos were dissociated from one another and placed singly or in groups, either internally or externally, on carrier embryos. The fate of the donor cells, marked by prior labelling of their nuclear DNA with [^{3}H]thymidine or with GPI-isoenzyme difference, was then followed in subsequent development. The results showed a clear biasing of dell fate toward either TE or ICM development as a function of donor cell positioning on the recipient embryos. Cells placed externally contributed over 90% of their descendants to the outside region of the blastocyst, while internal placement produced a less dramatic but still substantial (40%) contribution of cellular descendants to the ICM. Furthermore, when 8-to 16-cell embryos were enclosed by six unlabeled embryos, aggregates were formed that developed into larger blastocysts; of these several showed labeling only within the inner cells and developed into normal appearing embryos by 13 days. In these embryos, cells that would have developed into TE had evidently been channeled by their position into forming ICM. Finally, chimeras with externally placed cells that were allowed to develop to birth showed reduced contributions from the donor cells to the development of the coat, as scored by coat colour markers, regardless of their specific genotype. This result is the expected one if external cells are preferentially channeled toward TE formation and away from ICM formation. In sum, the results indicate that position rather than intrinsic blastomere cytoplasmic inheritance is a primary factor in influencing blastomere fate.

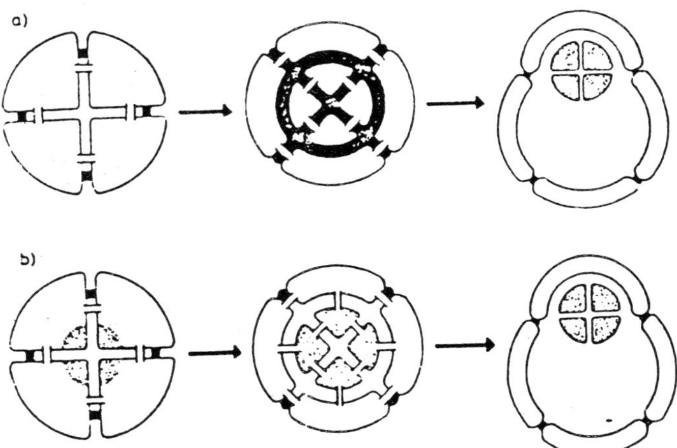

Figure 9.6 : Two models of positional fate setting in the mouse embryo. (a) The permeability block microenvironmental hypothesis; (b) the polarization hypothesis.

How might cellular position translate into a molecular mechanism for the differential restriction of cell fate? The first explanation proposed was that of *Ducibella (1977),* who suggested that inside cells find themselves within a qualitatively distinct microenvironment, created by the formation of apical zonular tight junctions between the TE cells that seal off the ICM cells completely from the external medium. The unique microenvironment of the internal cells then leads, by a series of steps, to the final differentiative and determinative steps.

However, two facts tell against this explanation. First, zonular tight junctions only develop at about the 32-cell stage *after* the initial divergence of the two cell types at the molecular level. Therefore, the first events in divergence take place without a communications barrier between the two cell types. Second, it has so far provent impossible to create by artificial means the postulated microenvironment. One observation, in particular, is fairly striking: injection of an entire eight-cell embryo into a fully expanded blastocyst failed to prevent the outer cells of this enclosed embryo from turning into TE despite the tight permeability block to the external environment provided by the host blastocyst.

The microenvironment hypothesis of Ducibella might be termed an extreme "instructionist" one : the environmental difference is a complete, qualitative difference that automatically produces a corresponding qualitative cellular difference. However, if the initial environments of outside and inside cells are not different in kind but only in degree, the initial cellular differences may also be subtle ones of degree. In slightly different terms, the initial difference may be quantitative, involving a graded property of some kind.

Johnson et al. (1981) proposed a hypothesis along these lines. Their "polarization hypothesis" stated that inside-outside differences stem from a radial asymmetry within the embryo in which individual cells are polarized with respect to molecular and cellular properties at their (outer) apices and (inner) bases. In this view, cleavage in the compacted eight-cell embryo, occurring perpendicularly to the radial axis, separates cells with "inner" (presumptive ICM) and "outer" (presumptive TE) values.

This hypothesis makes compaction a key event in creating or amplifying a radially-graded property. Compaction also draws the eight cells together into an organized unit in which communication is maintained through extensive intercellular contact and small intercellular channels at gap junctions. The importance of compaction in the separation of ICM and TE has been demonstrated experimentally using treatments that reversibly inhibit compaction during development. With one treatment that inhibits compaction without affecting cleavage, a delay in compaction beyond a certain time produces blastocysts that lack a genuine ICM component, yet the cells of these aberrant blastocysts synthesize the full spectrum of proteins characteristic of normal blastocysts. Under these conditions, the basic gene expression program is activated, but in the absence of cell compaction, none of the cells evidently perceive themselves as ICM.

An important second set of observations that support the hypothesis concerns the existence of individual cell polarization at the eight-cell stage. Whether in the intact embryo or in isolated cell pairs, cells at this stage become columnar and axially polarized, showing an outer cap of microvilous extensions that act as binding sites for a fluorescent ligand (fluorescein isothiocynate-concanavalin A). When the cells of the eightcell embryo divide in vitro, most divisions generate a polarized cell (possessing the ligand binding sites) and a non-polarized cell (lacking these sites): the former correspond to outer cells, the latter to inner cells. Some 2/16 cell couplets (pairs of cells at the 16-cell stage) of this kind. In the normal embryo, these two types of cells tend to form two discrete lineages, polar cells dividing to give pairs of polar cells and apolar cells dividing to give apolar cells. The beginnings of the ICM-TE lineage divergences are foreshadowed in these cytologically defined lineages.

Doe's the suggestion of a radial gradient conflict with the discovery that cell fate can be biased in one direction or the other by the appropriate changes of position within the early embryo? There is no essential conflict if the postulated gradient, with its high point at the center and decreasing along all axes out to the periphery, can "entrain" cells by means of cell-cell communication to assume the characteristic gradient value at any position.. *Hillman et al.* (1992) point out that their results are consistent with a gradient hypothesis if the gradient source exists at the center of the embryo; transplanted blastomeres would tend to acquire the radial gradient value of their new position.

However, whatever the origins of polarity characteristics, whether they arise in the embryo or preexist in the oocyte, the maintenance of the two cellular lineages requires continuing interactions between polar and apolar cells. While isolated 2/6 (polar :apolar) couplets give rise predominantly to four-cell (4/32) clusters containing two polar and two apolar cells, synthetic couplets consisting of two polar (TE-like) cells produce some ICM-like cells following division. The results indicate that it is the continuing interactions between polar and apolar cells that tend to maintain the two cell type lineages.

The results indicate a complex origin of the "morphogenetic' instructions" in the early mouse embryo involving both preexisting organization and cell-cell interaction. Position is important in fate setting, as shown by the experiments of *Hillman et al.* (1992), but it is position with respect to the site of origin within the oocyte mass and to the other cells of the early embryo rather than to the external environment that is a primary importance.

There are two other features of the ICM-TE divergence that deserve note. The first is that there may be certain regularities in cleavage that affect cell fates. From a study of cleavage rates and spatial patterns in partially flattened embryos in vitro, C.F. Graham and his colleagues conclude that at the two-cell stage, one blastomere always cleaves first and its progeny tend to occupy internal positions preferentially. If the observations pertain to normal embryos, they suggest that a topogenetic bias leads to a biasing of cell fates. The other observation is that the first inner cells show more rapid cycles of nuclear replication than the outer cells. In this system, as perhaps in others, it appears that cells with greater range of developmental potential show more_ id rates of nuclear DNA replication.

The origins of the second pair of determinative events in the preimplantation embryo are more obscure than the ICMTE divergence. However, they presumably also involve the intrinsic properties of the progenitor cells (established by the earlier events) and the interaction of these cells with each other and with the environment. The divergence of the two kinds of TE cells-polar and mural-seems to depend, as mentioned earlier, on the special interaction of polar cells with the ICM mass. In the absence of that interaction. TE cell invariably become mural TE. The divergence of endoderm from ectoderm in the ICM may, however, depend primarily on different relative exposures to the blastocoelic fluid. When ICM cells are placed into empty zonae, they invariably develop into endodermal cells. In intact embryos, the embryonic ectoderm, enclosed between endoderm and polar TE. remains totipotent and is the ultimate source of the fetus.

In both the TE-ICM and the endoderm-ectoderm divergence,

it is the relatively enclosed inner cell group that retains the undifferentiated appearance of early morula cells and the grater set of prospective potencies. The high point of developmental capacity in the preimplantation embryo is thus at or near the center; cells formed from the periphery differentiate first and exhibit a smaller range of fates.

POSTIMPLANTATION DEVELOPMENT IN MOUSE

In appearance, a chicken egg and a mouse egg are very different. Most obviously, they differ vastly in size, roughly 5-6 cm average diameter in the former compared to 80 Kin in the latter. Despite this difference, the basic embryology of the chick and the mouse have much in common. The difference in egg size reflects the different ways of provision for the nutrition of the developing embryo. The chick embryo has all of the nutrients that it needs prepackaged into the egg, in the form of yolk and albumin, while the mouse embryo must derive virtually all of the material and energy it needs for its growth and development directly from its mother. The mammalian arrangement requires that the embryo construct a much more elaborate set of extraembryonic membranes than the chick, membranes that will house it and connect it to its maternal parent.

Nevertheless, the relative simplicity of chick embryogenesis provides a suitable introduction to the mouse and will be briefly described here. The unfertilized chicken egg consists of a large mass of yolk capped by a small disc of cytoplasm containing th egg nucleus. Fertilization of the membraneenclosed egg occurs the egg passes down the oviduct, before the hard outer shell formed, and it succeeded by rapid cleavage divisions, wh generate a disc of approximately 80,000 cells by the time of la This cell layer or blastoderm consists of a relatively thin c area (the *"area pellucida"*) and a thicker outer ring (the *opaca"*). At one side of the area pellucida, a secondary thi about five cells deep is apparent. This thickening is te "primitive streak" and constitutes the direct embryonic region. By a series of cell movements, the primitive s comes to be composed of two cell layers. The upper like layer is termed the *"epiblast"* and the lower, more endodermlike layer the *"hypoblast,"* The embryo itself develops exclusively from the epiblast, while the hypoblast gives rise to the major extratmbryonic membranes of the yolk sac. These membranes separate the embryo from the yolk and serve as a conduit for nutrients from the yolk to the developing chick. The cell movements that generate and separate the epiblast from the hypoblast are complex and constitute the process of gastrulation; they involve migration of cells toward and through the primitive streak area.

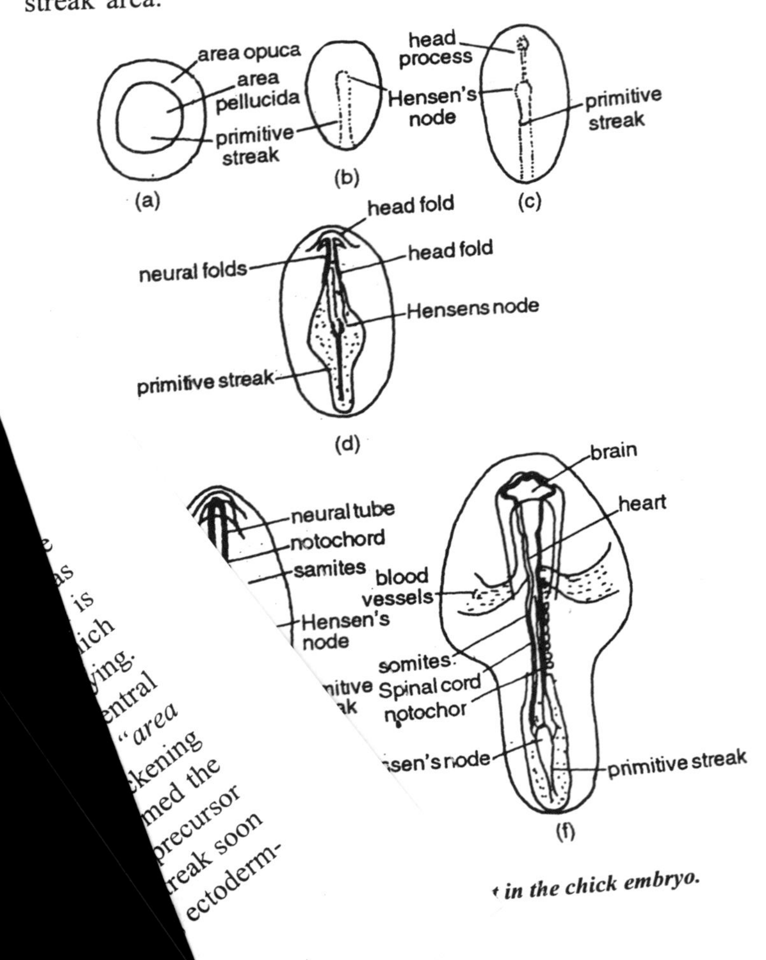

t in the chick embryo.

The primitive streak defines the future a-p axis of the embryo. The next major phase of development involves the creation of the major longitudinal (axial) sequence of structures in the embryo. The primitive streak lengthens, and this lengthening if followed by the formation of a condensation at the anterior end, to form the structure known as "Hensen's node." Hensen's node is mesodermal and gives rise to the main mesodermal elements of the early embryo. A portion elongates in the anterior direction, underneath the outer ectodermal cells, and thereby lays down the precursor cells of the notochord, a transitory central element in vertebrates. The most anterior point of this migration marks the position of the "head process," the feature cephalic region. The remaining portion of Hensen's node migrates posteriorly, under that portion of the ectoderm that gives rise to the neural plate, the precursor of the spinal cord. This migration is accompanied by the delimitation of the somites, which are paired mesodermal, segmental units. The somites are the ultimate source of the axial skeleton and ribs, the body musculature, and the major part of the dermis of the adult bird. As somite formation progresses, the neural plate closes to form the neural tube. The ensuing developmental events in the chick embryo are so closely correlated with the stages of somitogenesis that chick embryos can be reliably staged by reference to the number of somite pairs that have formed at any given time point. By the end of somite formation, the body plan of all of the major organ systems in the chick is discernible.

The mouse embryo also develops from the primitive streak, but the preamble to its formation is . considerably more complicated. At the time of implantation, which occurs in the mouse at about 4.5 days postoconceptus (p.c.), the balstocyst consists of just four cell types: the polar and mural TE cells and the primary ectoderm and primary endoderm. A full 2 days elapse during implantation before a detectable primitive streak arises. The complete sequence of events from implantation through the first stages of organ formation has been described by Snell and Stevens (1996), and is summarized here.

At the time of implantation, the internal endoderm covers the ectodermal ICM cells but has not spread around the inside of the blastocoele. During the next 1.5 days, while the embryo is embedding itself in the uterine wall, the primary endodermal layer comes to cover both the inside of the blastocoele and the inner ectodermal cell mass. The latter, covered with a stocking of endodermal cells, projects progressively farther into the blastocoelic cavity, and the whole structure is now referred to as the "egg cylinder"; the blastocoele at this point is designated the "yolk cavity" (which in the avian embryo is filled with yolk).

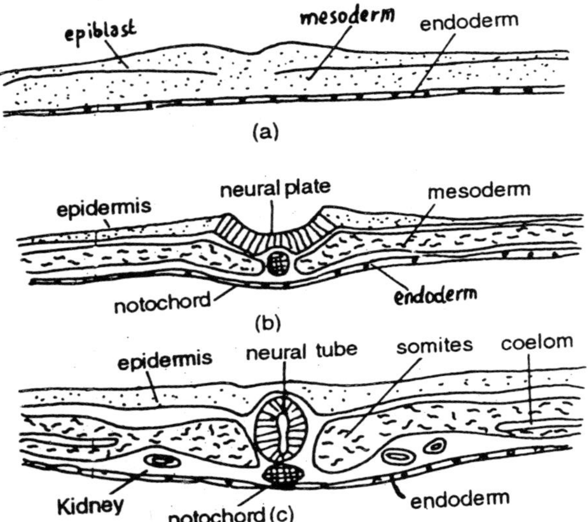

Figure 9.8 : Development of the chick neural tube. (a) Primitive streak stage; (b) neural plate development; (c) neural tube stage.

(This early invagination of embryo precursor material into the yolk cavity is peculiar to mice and rats, and does not take place in other placental mammals, even in more primitive rodents. The typical mammalian pattern is initially similar to the avian one; the ectoderm is a disc sitting atop the endodermal layer and invaginates only at a much later stage. The murine arrangement, in which ectoderm comes to be enclosed by endoderm, is referred to in classical embryological texts as the "inversion of the germ layers.")

The primary endoderm now consists of two distinct cells layers. The layer that covers the inside of the yolk cavity is referred to as the *"distal"* (or *"parietal")* endoderm., while that covering the egg cylinder is termed the *"proximal"* (or *"visceral")* endoderm. The proximal endoderm blankets both the embryonic ectoderm-so called because it is the precursor of the embryo proper-and the extraembryonic cells; the latter are derived from the polar TE cells, which retain their capacity for cell division. Capping the extraembryonic ectoderm in the 5.0-day embryo is the ectoplacental cone, also derived from the polar TE and a major precursor of the placenta. At this stage, a small cavity within the embryonic ectoderm can be. discerned, the proamniotic cavity, which is the precursor of the amniotic cavity, the fluid-filled sac containing the fetus.

The entire structure enclosed by the visceral endoderm is called the *"yolk sac,"* and from this point on, the visceral endoderm can be equated with the hypoblast of the chick embryo. The yolk sac enlarges within the yolk cavity, and by 5.5 days the embryonic ectoderm has begun to delaminatethe second of the germ- layers of the embryo, the mesoderm. Concurrently, the cavities within the embryonic and extraembryonic portions of the egg cylinder have enlarged and temporarily joined. Cavities have also begun to appear within the mesoderm; these cavities unite to form the beginnings of the exocoelom, which by 7.25 days occupies the space between the ectoplacental and amniotic cavities. The exocoelom is separated from the amniotic cavity by the amnion, a membrane composed of a layer of mesoderm (on the exocoelomic side) and layer of ectoderm. By 8.5 days, the exocoelom has surrounded the amniotic except for the region occupied by the fetus. By this stage, the ectoplacental cone has fused with the chorion (the membrane bounding the exocoelom opposite the amnion) and the allantois, a structure derived from the mesoderm that serves as a connection between the maternal blood supply and the fetus. The chorioallantoic membrane is connected to the fetus through the richly vascularized membrane that bounds the exocoelom, the yolk sac splanchnopleure.

The fetus originates at 6.5 days, with the appearance of the primitive streak as a thickened region on one side of the embryonic ectoderm, near the developing allantois. Between 6.5 and 10 days, all of the major organ systems take shape within the primitive streak through a complex series of morphogenetic movements accompanied by extensive growth. The process beings with the production and proliferation of mesodermal progenitor cells at the proximal (allantoic) end of the primitive streak; this position marks the future caudal end of the fetus. As ectodermal cells migrate through the primitive streak, they move both laterally and distally toward the future cranial end of the embryo. This migration of cells into and through the primitive streak constitutes gastrulation in the mouse embryo. The a-p orientation of the embryo is thus established or revealed by the laying down of the mesoderm in a caudocranial direction. As this process continues, mesodermal cells progressively occupy the space between the visceral endoderm and the embryonic ectoderm, except for the narrow central section that forms the spinal region of the fetus; this early primitive streak stage is very similar to that of the chick.

At 7.0 days p.c., the head process has taken shape at the most distal end of the egg cylinder. As in the chick, the head process gives rise to the notochord; it also contributes to part of the endodermal lining of the gut. The head process grows both proximally, on the side of the egg cylinder opposite the primitive streak, and laterally. By 7.5 days, the growing head process and the mesodermal sheets have met; throughout most of the embryo, ectoderm is separated from primary endoderm by the mesodermal layer, except for the midsaggital strip of head process directly in contact with the ectoderm.

The early primitive streak stage is a period of exceptionally rapid growth and transformation, During this period, the longitudinal elements, namely, the digestive tract and the axial system, begin to form. The gut begins as two invaginations of the yolk, sac on opposite sides, near the junctures of the embryonic and extraembryonic ectoderm. The foregut precursor pushes in at the cranial end, beginning at 7.0 days, and 7.75 days has become a pronounced inpocketing: the hindgut has just begun to form at this point. The blind ends of the foregut and hindgut invaginations eventually break through to the outer surface of the embryo, giving rise to the mouth and anus, respectively. The invagination of the foregut also lays the foundation of two other key organs, the heart and the brain. The most dorsal region of ectoderm that is pushed in is the head fold structure, which develops into the brain. The heart arises from the mesoderm lying between the head fold and the outer endoderm.

The axial system comes into being from the lateral sheets of mesoderm that border the notochord, the so-called paraxial mesoderm. As in the chick embryo, the axial system arises in a strict, linearized sequence of somite formation that begins at the cranial end of the embryo, just posterior to the embryonic brain, and progresses to the caudal end. Somitogenesis therefore proceeds in the opposite direction to that of mesoderm formation. The first pair of somites arise at about 8 days. Initially, pairs of somites are laid down at the rate of one per hour; subsequently, the process slows to one pair every 2-3 hr. Somitogenesis is complete by about 13 days of development, when 65 pairs of somites have been formed; of these, 30 pairs will form of the skeleton, musculature, and dermis of the body (neck, thorax, and lower back) and 35 pairs will form these elements in the tail. Each newly formed somite shows a trace of a boundary between the anterior and posterior halves, and each vertebra arises from cells of the posterior half of one somite and the anterior half of the next. Differentiation of the somites is accompanied by dispersal of the mesenchymal cells that compose them, leading to an obliteration of the intersomitic boundaries.

The stages of neural development, the precursor of the spinal cord and the other major component of the axial system, are closely correlated with the stages of somitogenesis. The neural tube develops from the closure of the neural plate area just above the notochord and between the sheets of paraxial mesoderm. The strict correlation of the somite stage with other events in normal development permits one to stage embryos according to their somite number, as in avian embryos.

By 10 days p.c., the major elements of the circulatory system, also formed from the paraxial mesoderm, have developed. The turning of the embryo also completes the formation of the gut by generating the midgut and joining the fore-and hindguts. During the succeeding 10 days of development in utero, the major and minor organ systems undergo progressive differentiation. A photograph of a *26*somite embryo and a diagram of the 13.5 day embryo. A timetable of some of the major events in postimplantation development is given in Table.

Growth Size and Morphogenesis

One of the most general and remarkable features of embryogenesis is the co-ordination of growth with the events of development. This co-ordination ensures that the newborn or newly hatched progeny of a given species or strain are of a relatively uniform size and that the internal and external structures are correctly proportioned.

The existence of mechanisms for co-ordinating development with growth becomes obvious when embryo size is altered experimentally. The most dramatic of such regulatory responses are those seen when isolated blastomeres from early embryos give rise to whole animals. For instance, if one separates the blastomeres of the four-cell sea urchin embryo, each cell develops into a perfectly shaped and normally functioning pluteus larva, although only one-quarter the normal size and containing one-quarter the normal number of cells. Similar regulative capacities are displayed by the first blastomeres of frogs and starfish; in mammals, normal newborn rabbits have been obtained from isolated blastomeres of eight-cell embryos.

TABLE 9.1 : DEVELOPMENTAL LANDMARKS IN POSTIMPLANTATION DEVELOPMENT

Event	*Days p.c.*
Implantation	4.5
Proamniotic cavity appears	5.0–5.5
Primitive streak detectable	6.5
Foregut appears	7.0
Neural plate forms	7.75
Somite formation begins	8.0

Chorioallantoic placenta forms	9.0–10.0
Primordial germ cells appear at the base of the allantois	7.75–8.0
Primordial germ cells arrive in genital ridges	10.5–11.5
Melanoblasts leave the neural crest	8.5–9.5
Neural tube development completed	10.0
Forelimb buds appear	9.2
Hindlimb buds appear	10.0
Somite formation completed	13.0
Hematopoiesis begins in the fetal liver	9.5–10.0
Hematopoiesis begins in the fetal bone marrow	16.17
Birth	19.20

**Times cited are approximate; inbred mouse stains frequently show a delayed timetable.*

Major adjustments can also occur much later in development. Individual duck blastoderms, for instance, if operated on prior to the emergence of the primitive streak, can give rise to four independent embryos. In *Drosophila,* severe limitations of the food supply during the period of larval growth can lead to the production of miniature, although otherwise fully normal, fruit flies.

In the mouse, growth regulation can occur at both early and intermediate points of development. Chimeric blastocysts made from two or more embryos are invariably larger than normal blastocysts, yet develop into newborn mice of standard size. At some point, there must be a "downward" regulation of size to yield a normal-sized embryo. The point has been localized to between 5.5 and 6.0 days days p.c.; it occurs, therefore, after implantation but before primitive streak formation. This response probably reflects a limitation of the nutrient supply to the larger blastocystic embryos, with a consequent slowing down of cell division rate.

"Upward" regulation of the growth of abnormally small embryos also occurs, although after primitive streak formation. Unlike the sea urchin embryo, viable mammalian embryos formed from single blastomeres of the two and four-cell stage do not give rise to one-half or one quarter size animals, but to newborns of normal size. The size adjustment apparently occurs between days 10.0 and 11.0 p.c., and takes place shortly after the maternal circulation is made fully available to the embryo, with complete development of the placenta. The nutrient supply may thus make possible the growth spurt, but the signal to do so must come from the embryo itself, which somehow "knows" that it is small.

The most striking of upward regulation of growth occurs following treatment with mitomycin C (MMC). This drug, when injected intraperitoneally into mothers at the beginning of primitive streak development, produces extensive random cell death in the embryo. At 7.5 days p.c., the epiblast of MMC-exposed embryos contains only 10-15% as many cells as that of control embryos, the dead cells being sloughed off.

Despite this extraordinarily large reduction in cell number, the majority of embryos manage to catch up to control embryo size by 12.0-13.5 days, and many are born alive. The primary defect in the survivors is a partial or complete sterility, reflecting a permanent reduction in germ cell number effected by the MMC treatment.

Examination of MMC-treated embryos shows that the developmental time course of both ectodermal and mesodermal components is affected, with a visible retardation of neural tube development and of somitogenesis. However, the effects on the neural tube and the somites are partially uncoupled, with the neural structures experiencing the more severe delay.

It follows that the stages of development of the two major components of the axial system are neither obligatorily coupled to, nor dependent on, one another during the period of recovery (nor, by inference, during normal development).

An intriguing aspect of somite development in MMCtreated embryos is that while the normal number of somites is produced in the catchup embryos, the individual somites are smaller and have fewer cells (Tam, 1981). The conservation of somite number at the expense of individual somite cell content suggests that there exists some form of global partitioning mechanism within the presomitic mesoderm. This mechanism operates prior to the first segmentation event and ensures that the total mass is divided into the correct number of somite pairs. The existence of such a pattern-setting mechanism is, at least superficially, reminiscent of the partitioning of the *Drosophila* blastoderm surface into segments, although somitogenesis appears to have a much greater regulative capacity (and operates with greater cell numbers).

Gene Expression in Postimplantation Development

The events of implantation mark a point between the early but limited cell diversification of preimplantation development and a much more extensive set of cellular differentiations. Is this acceleration in cell diversification accompanied or driven by a corresponding set of qualitative changes in the gene expression pattern? The extensive evidence for *Drosophila-* and the less complete data for *Caenorhabditis* indicate that successive developmental stages and different cell types exhibit a large set of conjointly expressed genes, with comparatively few qualitative differences between these gene sets. Is the mouse similar or different in this respect?

Answers to this question have been sought primarily by protein labeling and electrophoresis techniques and by RNADNA hybridization techniques. The results indicate that, as in the fruit fly and the nematode, there is a broad base of shared gene expression between different cell types and stages during mouse development. At the same time, they show that largescale quantitative shifts in gene expression or gene product accumulation occur during development.

Labelled Polypeptide Patterns

The first developmental events in the postimplantation embryo involve the elaboration of the principal extraembryonic membranes and the development of the primitive streak. By 7.5 days of development, the polar TE has given rise to distinct ectoplacental cone (EpC) region, a layer of giant cells (GCs) that surround the diploid core cells of the EpC, and a layer of extraembryonic ectoderm (EE), which includes the chorion. The future embryonic region is separated from the exocoetom by the amnion and consists of the region of the primitive streak and, opposite it, a region primarily composed of embryonic ectoderm (EmE).

Johnson and Rossant (1981) assayed the molecular correlates of this first stage of postimplantation change by comparing the polypeptide labeling patterns of the three polar TB lineages with each other and with embryonic ectoderm. The different regions were dissected, labeled in vitro, and the polypeptides then extracted and separated on two-dimensional gels. In a large number of such comparisons, 50 spots, from a total set of more than 500 detected, were found that distinguished the four cell types from one another. Of these, six were unique to EmE, one.to EpC, and three to GCs. All of the remaining polypeptides in this set of 50 were shared between different subsets of the four tissues. Although the comparisons themselves provide no direct information about developmental relationships between the tissues, an in vitro culture experiment showed that cultured EE cells first develop an EpC-type profile that is then succeeded by a GC-type profile. By this criterion, the EE cells may be equivalent to the original polar TE cells and serve as a reservoir of EpC and GCs. The principal finding is that within the four ectodermal cell groups compared, which differ in cytology, function, and fate, there is a large degree of similarity in the detectable polypeptide synthesis pattern.

Later periods of development show much the same similarity of proteins synthesis. A detailed study of polypeptide synthetic patterns in embryonic mouse organs, employing [S] methionine pulse labeling in vitro, was reported by *van Blerkom et al. (1982).* The organ systems tested included liver, brain, kidney, forelimb, hindlimb, yolk sac, lung, and fetal tail (essentially somites) from 10- to 13-day embryos. For all stages and all organs, a total of 850-1000 polypeptide spots were detected. For the various organs surveyed, an average of three organ-specific protein synthetic differences were found, or less than 1 % of the total. In comparison, quantitative shifts specific to organs or stages were relatively more common. In the lung, for example, 5 qualitative changes in spot pattern and 15 quantitative ones were measured in the period from 10 days p.c., to early postnatal development.

For the brain, from day 10 p.c. to day 10 postnatally,1 qualitative change and 20 quantitative changes were detected. The very small number of observed changes in this survey was not a reflection of limitations in the labeling or electrophoretic procedures; labeling with ^{14}C amino acids or changing the pH range in the separation increased the total number of detected spots without altering the percentages of qualitative change in pattern.

The results of this and other studies not cited here are all in general agreement: there is little difference in polypepeptide labeling patterns between cells that are markedly different in phenotype. The two-dimensional separation procedure, however, may only resolve the various housekeeping proteins, all of which should be present in comparable amounts between different cell types. However, not all housekeeping proteins are necessarily in the abundant or semiabundant classes. *Galau et al.* (1976) have presented arguments that physiologically significant metabolic functions can be carried out by proteins produced from the rare mRNA class. Conversely, some of the detectable polypeptide spots may be produced from relatively rare mRNAs if translation efficiencies differ between certain species. Nevertheless, whatever the cellular functions of the detected proteins or the relative abundances of their mRN ks, they must represent only a portion of the expressed proteincoding genes. In contrast, RNA-DNA hybridization methods provide a more inclusive measure of the complete set of the number of expressed genes.

mRNA Pol Studies

Several measurements of mRNA sequence diversity in mouse cells have been made. Sequence complexity was estimated in these studies from the kinetics of hybridization of poly A^+ mRNA to cDNA made from such RNAs; the advantages of this approach were described earlier. The chief disadvantage is that only minimal estimates of sequence diversity are obtained. The true sequence diversity is underestimated in two wlys. Firstly, any mRNA species that are transported to the cytoplasm without poly A tails will not be purified; hence, these sequences will be missing from the cDNA preparation. In *Drosophila*, at least, these molecular species apparently comprise a substantial portion of the total mRNA pool. Secondly, rare mRNAs that are present in some but not all cell classes will not be detected because of dilution in the total mRNA pool. Nevertheless, if the presence or absence of polyadenylation is characteristic of a particular mRNA sequence, the measurement of sequence complexities in different cellular poly A mRNA pools will provide meaningful comparisons between different cell types. The three studies described below show that most mouse tissues have approximately 10,000 poly A mRNA species, of which the great majority are shared between cells of very different types.

In a comparison of such mRNAs from adult liver and brain and from whole 14-day embryos. *Young et al.* (1996) found that the three mRNA pools are complementary to 0.7, 1.7, and 0.7% of the sequence of the mouse genome, respectively. If one takes 20000 b.p. as a typical mRNA length,. these percentages correspond to approximately 12,000 different mRNA species for the liver, 25,000 for the brain, and 10,000 for the 14day embryo. (Although the brain is well-developed in 14-day embryos, the lower sequence complexity for whole embryos relative to the brain presumably reflects the dilution of rare brain-specific sequences in the former.) By performing crosshybridizations to near saturation in order to detect sequence overlap, Young et al. found a large measure of qualitative similarity between the mRNA pools of adult liver and brain. However, the rates of cross-hybridization, which primarily reflect the more abundant species, were lower than those in the homologous hybridizations. The measurements therefore indicate that the most abundant species for each of the two tissues were less prevalent in the other,

A similar analysis *by Hastie* and *Bishop* (1996) of mRNA populations of adult liver, brain, and kidney produced estimates of 13,000,11,600 and 11,500 different structural gene sequences in the three messenger pools, respectively. (The distinctly higher sequence complexity of the brain mRNA pool reported in the previous study was not seen in this one; the reason for the discrepancy is unclear). For all three tissues, three discrete abundance classes could be distinguished, consisting of the very abundant species (4-9 types), the intermediate (480-750), and the rare (11,000-12,200). Of the rare 11,000 poly A$^+$ mRNA sequences of the kidney, 9000-10,000 were estimated as shard by liver and brain. As in the study of Young et al., the abundant mRNA species for each tissue were also detected in the mRNA pools of the other two but were in the intermediate or rare class. Altogether, a total of 15,000-16,000 mRNA species are found within the three tissues. Each tissue sampled in the analysis represents one of the major germ layers of the fetusthe liver, endoderm; the kidney, mesoderm; the brain, embryonic ectoderm-yet the vast bulk of the structural genes expressed as poly A$^+$ messengers are shared between them. The main qualitative differences are in the rare mRNA classes, although quantitative differences of the abundant and intermediate classes are a marked feature.

Ideally, one would like to trace the entire pattern of transcript of diversification that occurs from the beginning of fetal development, in the primitive streak stage, on. Unfortunately, this has not yet proven possible because of the difficulty in obtaining enough early embryonic material. An alternative approach is to use neoplastic cells whose characteristics mimic those of early embryonic cells. These cells are termed "embryonal carcinoma (EC) cells" and have been widely used as an analogue of the early embryonic cellular condition.

EC lines come in at least four grades of developmental capacity, ranging from those that give rise to all cell types (totipotent lines) to those with wide but incomplete capacities *(pluripotent lines)* to those incapable of giving rise to any differentiated cells (nullipotent lines). *(Nullipotency is* associated with long-term culture and accumulated chromosomal abnormalities). A few lines have specialized differentiative capacities, such as a particular capacity to give rise to muscle cells under a differentiative stimulus. The different grades of developmental capacity may provide analogues of the various stages of embryonic cell specialization.

If toti- or pluripotent EC cells are truly similar to early embryonic cells, then a measurement of transcript diversity in EC cell lines should give a hint of the gene expression pattern in ICM or primitive ectoderm cells. *Affara* et al. (1997) measured EC mRNA transcript diversity in two pluripotent cell lines and two specialized, EC-derived myoblast (muscle precursor) cell lines. For the pluripotent EC lines, a transcript diversity of 7700 different poly A' mRNAs was estimated, and for the myoblast lines a transcript complexity equivalent to 13,200 poly A* mRNA species. By cross-hybridization tests, all of the EC mRNA sequences were found in the transcript pools of the myoblast cell lines, but the latter cells contained a substantial number of rare mRNA species not found in the EC cell line. As before, the abundant and intermediate abundant sequence classes that characterize one cell type are less abundant in other cell types.

Taking EC cells as an analogue of early embryonic cells, the comparison, suggests that divergence of cells from the early embryonic cell state is accompanied by an expansion of gene expression, principally in the class of mRNAs present in just a few copies per cell. However, other factors, such as the different origins of the various cell lines, may have contributed to the differences between the EC and myoblast cell lines.

Another investigation showed that terminal differentiation is accompanied by relatively little change in overall transcript diversity. The cells of a Friend erytholeukemic line-a virally transformed cell line with the potentiality to differentiate into red blood cells-can be induced to differentiate into mature red blood cells by exposure to the permeabilizing agent dimethyl sulfoxide (DMSO). When these cells, whose sequences are all common to EC cells apart from a few in the most abundant class are induced to differentiate, the principal changes occur in the most abundant species. These changes include a decline in the frequency of several sequence types and the appearance of new ones, including, most characteristically, the hemoglobin mRNAs. Thus, neither commitment to a given pathway of differentiation nor the process of overt differentiation itself necessarily entails large-scale changes in overall gene transcript diversity relative to the (presumptive) early embryonic cell state. Whether the biologically significant changes are primarily those in the abundant classes or whether changes in all classes are equally important remains an open question. However, in all specialized cells whose physiological function depends on a certain protein or group of proteins, one change will always be an increase in the concentration of the abundant mRNA(s) that code for those protein(s).

Changes in cellular mRNA concentrations may occur either through regulation of transcription or through changing efficiencies of transcript processing. There is some evidence that control of processing plays some role in regulating mRNA abundances. When the nuclear RNA pools of different mouse cell types whose cytoplasmic mRNA compositions differ are examined, they are found to contain sequences for the same rare mRNA types. Thus, the nuclei of brain cells have copies of all of the sequences present as mRNA in the cytoplasm of kidney cells and vice versa. However, the nuclear sequences that do not become converted to mRNAs are rare, being present at only one or a few copies per nucleus. These results indicate that there,may be some transcription of all potential mRNA sequences in all cells.

On the other hand, the regulation of the more abundant species may be subject to specific transcriptional control. Derman et al. (1981) cloned several sequences from cDNA made from liver poly A^+ mRNA, and identified 11 cDNA clones corresponding to abundant or semiabundant mRNAs that are much rarer in two other cell types that were examined-hepatoma cells and mouse L cells. For all 11 of these "liver-specific" sequences, the rate of pulse labeling of these sequences in nuclei in vitro-which serves as a measure of the instantaneous transcription rare-was found to be significantly higher in liver nuclei than in brain nuclei. In contrast, the in vitro pulse rate labeling of mRNA complementary to five cDNA sequences shared by several cell types was found to be equivalent in liver and brain cell nuclei. The experiment indicates that the more abundant mRNA sequences in cells are more concentrated precisely because the transcription rate of their homologous gene sequences is specifically enhanced. Transcriptional regulation in eukaryotes may therefore involve primarily modulations in transcription rate that shift sequences between the more and less abundant categories, while the presence or absence of certain rare sequences as mRNAs may involve some degree of sequence-specific posttranscriptional processing.

POSTIMPLANTATION DEVELOPMENT : GENETIC ANALYSIS

Postimplantation development is a continuous sequence of events but may be divided arbitrarily into two phases. The first extends from implantation at 4.5 days to approximately 6.5 days p.c., during which the extraembryonic membranes and the primitive streak are formed. The second and longer phase, from 6.5 days to birth, is the period of organ rudiment formation, organogenesis, and fetal growth. Everything that precedes the primitive streak stage involves the elaboration of structures necessary for the housing and nutrition of the embryo proper that arise from the primitive streak.

A large part of contemporary murine developmental genetics has centered on the delineation of the various component cell lineages of the tissues and organs of the postimplantation embryo by means of genetic techniques. Because mitotic recombination is either rare or nonexistent in mammalian embryos, it cannot be used to reconstruct the detailed history of these lineages. Instead, principal reliance has had to be placed on the construction and analysis of chimeras. As in the study of preimplantation lineages, early embryo chimeras are first made either by morula aggregation or by injection of cells into blastocysts. The chimeric embryos are then implanted in pseudopregnant foster mothers and, following the requisite period development, the animals are dissected and the tissue or structure of interest is analyzed in terms of the input marker phenotypes. From the pattern of distribution of the marked cells, deductions can then be formulated about the nature and/or number of the founding cells. This form of analysis has proven quite successful for preimplantation development, as we have seen, and for early postimplantation development. For post-primitive streak development, the method has been applied principally to estimating the number of founder cells for different organs or tissues; in these studies, the interpretations, while interesting, have been consistently controversial. A second form of clonal analysis is that based on the inactivation of single X chromosomes that occurs during the early development of two-X (female) embryos; it has also been employed to estimate founder cell numbers. However, these analyses also suffer from ambiguities that affect the interpretation of the results. The findings from these approaches, and the difficulties, are reviewed below.

Origins of the Extraembryonic Membranes and Fetal Germ Layers

At implantation, the TE lineage of the early blastocyst has split into two distinct cell groups. The mural TE cells, which occupy the sides of the blastocyst, have begun the transformation to primary GCs. These cells eventually achieve ploidy levels of several hundred; their function is to achieve the initial implantation of the blastocyst in the uterine wall. The other TE cell group, the polar TE, produces the Epc and EE cells. The presumptive cell derivation relationships in the polar TE lineage wee described earlier; the EE cells may be the primary type and give rise to EpC cells, which in turn generate secondary GCs at the most proximal end of the embryo.

Although there has been little controversy about the TE lineages, the origins of at least two fetal tissues have been the subject of debate in the recent past. One of these tissues is the fetal endoderm. From the classic embryological description of the formation of the gut, it had been assumed that the proximal endodermal layer gave rise to the lining of the gut; without question, endodermal cells from this layer are carried along in the imaginations that give rise to the fore- and hindguts. However, this source is difficult to reconcile with the development of the avian embryo: in the latter, the entire embryo, including the gut, arises from the epiblast. The avian hypoblast, which appears similar in function and location to the proximal endoderm, does not contribute to the' embryo proper.

The origins of the germ line, the gamete-producing cells, have been the second contentious issue. Direct histological tracking of the primordial germ cells (PCGs) had placed their source, like that of the gut, in the proximal endoderm. A number of observations have forced a revision of this view; it is now apparent that the germ line, like all the other components of the fetus, comes from the EmE.

A feature of the PGCs that has made it possible to trace them in early development is their high level of alkaline phosphatase (AP) activity. By staining for AP activity, was able to detect presumptive PGCs in the 8.0-day p.c. embryo near the caudal end of the primitive streak, within the base of the allantois. Twelve hours later, the main cluster of high AP cells was observed in the proximal endoderm and, 12-24 hr later, within the invaginated hindgut rudiment. From this position, the PGCs then move by an active, chemotactic process, along the dorsal mesentery within the developing body cavity, to the genital ridges, which they then colonize. (This pattern of movement occurs in embryos of both sexes; all of the differences in germ cell maturation between the sexes begin to be apparent after colonization of genital ridges). The process of migration is complete by about 11 days p.c. From the point at which they first become distinguishable to the end of migration, the number of histochemically detectable PGCs increases from about 100 to over 1000.

Two genetic approaches establish that the early epiblast or primitive steak is the source of the PGCs. The first involves the phenomenon of X chromosome inactivation in female embryos. Although female mammals, like female fruit flies, have two X chromosomes to the male's single X, mammals do not use the fruit fly's system of transcriptional dosage compensation. Rather, in each somatic cell of the female embryo, one X chromosome is totally inactivated through chromosome condensation. The inactive state is then inherited by all of that somatic cell's progeny. This transcriptional inactivation happens after the blastocyst stage in all cells of the embryo. In the endodermal layer, inactivation occurs early, and it is always the paternal X chromosome that is inactivated. In the epiblast (the EmE and all its derivatives), inactivation is slightly later and is random, with half of the cells experiencing inactivation of their paternal chromosome and the other half that of their maternal X. If the germ cells were of endodermal origin, they should all show an active maternal X only. In fact, the PGCs within the genital ridges always show the epiblast pattern, a random inactivation of the maternal and paternal Xs.

Chimera experiments supply an even more direct proof that the germ line comes from the primitive ectoderm. These experiments not only pinpoint the source of the germ line but also establish that it does not arise from a defined germinal plasm. This fact marks a significant difference from the eggs of many insects, nematodes, and anuran amphibians, among other animals. The first experiments indicating that the germ line arises from non-specialized blastomeres were those of *Kelly* (1975). She found that in each of two four-cell embryos dissociated to individual blastomeres, at least three of donor cells gave rise to germ line chimeras. In all cases, the germ line chimeras also displayed mosaicism in their somatic tissues, showing the absence of segregation for germ line-forming ability by as late as the second cleavage division.

Even more conclusive are the experiments of Gardner (1977), who injected single ICM cells from 3.5 day p.c. blastocysts or single embryonic ectodermal cells from 4.5 day blastocysts and scored for subsequent somatic and germ line mosaicism. For the 3.5- and 4.5-day donors, there were, respectively, four out of six and two out of seven somatic chimeras that also showed germ line chimerism (as evidenced by progeny possessing the donor germ line marker). Although the number of chimeras in these experiments was not large, the frequencies were typical for single-cell injections. The experiments show that there is no specialized germ line determinant in the mammalian oocyte and that the primary ectoderm is the source of the germ line. In mammals, therefore, the continuity of the germ line is maintained through the soma rather than independently of it. The same is true of the avian embryo; reciprocal inteispecies chimeric hypoblast-epiblast combinations of quail and chic reveal an epiblast origin for the germ line.

The complete sequence of major tissue derivations in the mouse embryo, reconstructed primarily from chimera studies, is summarized. The chief conclusions are that all the cells of the fetus are derived from a portion of the EmE and that the placenta is composed of several discrete lineages of both the TE and the ICM.

Genetic Analysis of Tissue and Organ Foundation

The period of major tissue foundation in the mouse embryo occurs within 24-hr of the formation of the primitive streak. However, many of the details of histogenesis and organogenesis are obscure. The two aspects that have received most attention are the first cellular commitments in organogenesis—whether they constitute restrictions in developmental capacity (determination) or depend on cell movement-and the number of founder cells for each organ. One uncertainty that affects the analysis concerns the precise timing of the initial developmental assignments. Presumably, the first restrictions of developmental potential within the primitive streak occur between 65 days p.c. and the beginning of somitogenesis. By the late somite stages in avian embryos, distinct determinative states have been imposed; somites from the thoracic regions of the embryo generate vertebrae with ribs when transplanted to other regions of the embryo, but lumber somites, those from the lower back region, never form ribbed vertebrae. If one assumes that cells of the early primitive streak mouse embryo are unrestricted, as they are in the avain embryo, a considerable degree of narrowing in developmental potential occurs between the early and late primitive streak stage.

Experimental embryology offers one set of approaches to establishing the nature and timing of the first developmental commitments in primitive streak embryos. One method is to cut out and culture pieces of these embryos at specific stages and to see what they form. The range of structures that can be produced by a given piece from a given stage serves as a measure of the autonomous developmental capacities of that piece; if specific embryo fragments always produce specific characteristic structures, the fragments may be presumed to have experienced some assignment of developmental role. *Snow* (1981) has reported the results of such a study. Isolated pieces of early (7-day p.c.) to late (8.5 day p.c) primitive streak embryos and the complementary deficiency embryos were cultured in isolation for 24-hr in vitro and then examined for the structures formed. In these tests, examination of the deficiency embryo in conjunction with the fragment allows one to determine whether duplication or regeneration of structural occurs. A high frequency of differentiation (> 70%) of the cut pieces and an even higher frequency for the deficiency embryos (>90%) was observed.

These results can be related to the prospective fates of the different regions assayed. In all cases, the autonomous development of those tested pieces that produced recognizable structures corresponded to the prospective fate of that region in the unoperated primitive streak embryo. Thus, the most anterior third of the 7.5 day embryo, which is destined to give rise to the head fold, brain, and foregut, produced those structures in vitro. The middle third, which gives rise in vivo to the major mesodermal elements, was found to produce sometimes, neural tube, and some heart structures in vitro. Finally, the most posterior section, whose fate is to produce tailbud, hindgut, and PGCs, produced those structures plus allantois and no others. A further trisection of the posterior region permitted a mapping of the allantois to the most posterior section of the primitive streak, and the PGC-forming capacity to the section immediately anterior to this one.

What developmental property is mapped in this experiment? *Snow* (1981) has described the result as being that of an "allocation map," to indicate a preliminary regional localization of capabilities (without necessarily being fixed, determinative states). On the other hand, *Beddington* (1982) has argued that such an inference is invalid because the tested pieces are large and consist of more than one tissue type. In principle, the presence of a combination of cell types might establish or stabilize a range of specific inductions, each of which triggers a differentiative pathway whose end product is scored at the termination of the culture period. On the basis of the ability of late primitive streak ectodermal cells, labeled with [^{3}H]thymidine, the colonize different regions and germ layer derivatives, *Beddington* has suggested that the ectodermal cells at least are undetermined in late primitive streak embryos. A difficulty with this view is that colonization and replication of cells in a heterotopic site do not necessarily signify that the cells have taken on the developmental state of the cells with which they cohabit.

To date, therefore, the techniques of classical embryology have left unsolved the problem of when determination in the primitive streak embryo takes place. It may be, however, the before states of determination can be assessed, a fuller description of cell allocation in the early embryo is necessary. Allocation presumably precedes or accompanies determination and characterization of the timing and number of founder cells allocated to various tissues or organ primordia should provide a basis for the subsequent analysis of determination.

In principle, clonal analysis should be able to furnish information on the number of founder cells present at allocation from the estimates of numbers of descendant clones produced within the organ or tissue at the time of allocation. In the fruit fly, one first determines the approximate time of allocation by finding that point at which mitotically generated recombinant clones cease to mark neighbouring primordia (whose proximity can be judged from the gynandromorph fate map). Mitotic recombination is then induced at that state, and from the fraction of tissue or organ occupied by the marked cells, the number of founder cells can be calculated.

The mouse embryo demands a different strategy because marked cells cannot be induced by mitotic recombination in mammals; because one cannot generate marked clones at will during development, one cannot directly estimate the time of allocation in mammalian embryos. Rather, one must introduce the marked cell(s) at arbitrarily early times, either by direct physical placement to create chimeras or by the use of X chromosome inactivation in female embryos. If a female fetus is heterogygous for an X-linked gene-for instance, for a gene encoding electrophoretically distinct forms of an enzymehalf of its cells will express one form, the other half the other form. X inactivation therefore provides a natural form of genetic mosaicism within heterozygous female embryos, and thereby furnishes a second potential means of clonal analysis in the mouse embryo. By measuring the proportion of cells within in tissue that express one allele or the other, the experimenter can arrive at estimates of founder cell number. Unfortunately, as in chimera construction, the genetic heterogeneity is not under experimental control and occurs before primitive streak formation.

With either chimera construction or X inactivation, therefore, the genetic marketing of cells occurs before the presumptive times of fetal tissue' and organ allocation. As discussed earlier, the consequence is that marked structures tend to consist of disproportionately excessive amounts of marked tissue, producing systematic underestimation of founder cell number.

There is a further aspect of mammalian cell development, typical vertebrate development in general, that can complicate the analysis. In *Drosophila,* the descendants of single cells tend to remain contiguous; hence, most visible coherent clones are true individual descendant clones. In the mouse, this situation is much less true. Extensive cell mingling and cell migrations take place, particularly in later development. The result is that a given early clone will usually fragment into separate daughter subclones. If one mistakenly equates patches of marked cells with individual descendant clones, the estimate of primordial cell numbers will be inflated to a degree corresponding to the fragmentation of descendant clones. Furthermore, the estimate can be affected depending on whether cell migration occurs before or after tissue allocation, some of the effects of different patterns of migration and coherent clonal growth on final patch sizes and discreteness. When mingling is very extensive, individual patches may consist of two or more subclones that happen to be contiguous.

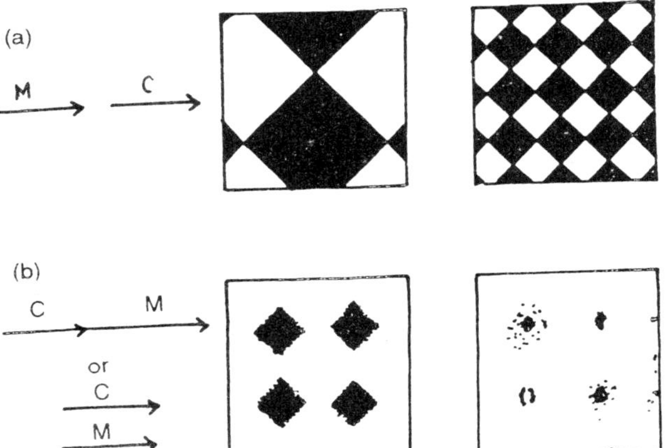

Figure 9.9 : Patch pattern as a function of the timing of coherent clonal growth and cell migration (a) Change in a pattern of large patches (left) produced by a period of migration (M) followed by coherent daughter clone growth (C) without subsequent migration; (b) patterns produced by clonal growth followed by migration or by a balance of clonal growth and migration.

The consequences and complications of all of these factors in estimating founder cell numbers in the mouse have been reviewed by *McLaren* (1986a) and *West* (1988). The ultimate result is that estimates from clonal analysis in the mouse are considerably less certain than the corresponding estimates in the fruit fly. Nevertheless, the mammalian studies have yielded minimum estimates of founder cell number and have illuminated the detailed patterns of cell growth in mammalian development.

Chimeras have been analyzed in two distinct ways to estimate founder cell numbers. One approach employs indirect cell markers —such as enzyme electrophoretic variants, which can be scored only in bulk preparations and involve statistical estimates of founder cell number from the proportions of the component genotypes in the chimeras. The second set of methods utilizes direct cell markers, which can be scored in situ by direct inspection or histochemical means. In these methods, the physical disposition of the marked cells within the total cell population provides clues to the number of founding cells or clones and to their growth pattern within the tissue.

The earliest approach to estimating founder numbers with the chimera technique was statistical. It was based on the assumption that formation of a tissue or organ primordium consisting of a fixed number of cells, can be likened to a lottery drawing in which cells are picked at random from a common precursor pool until the correct number has been drawn. In any large set of two embryo-aggregation chimeras, this method of primordium formation should generate a range of binomially distributed genotype compositions; the shape of the distribution is function of the number of founding cells.

Instances involving very small numbers of founder cells will illustrate the point. If a given tissue primordium is always started by one cell, then none of the samples of that tissue from a group of chimeras will be mixed composition; all will be genetically pure. If a primordium is always composed of two founder cells, then the fraction of chimeras that will have tissue of pure genotype a will be one-fourth (= 1 /2 × 1 /2) that of pure genotype b will also be ¼, and the fraction of chimeras that will have tissue will be one-half (= 2 × ½ × ½). By the same reasoning, if the cell number is three, then one-(fourth of the chimeras will be genetically pure (2 × 1 /8 for each genotype) and three-fourths will be mixed. Since, in practice, it is often easiest to score the fraction of embryos with non-chimeric tissues, some of the early estimates of founder cells number were based solely one the fraction of mixed embryos that were judged non-chimeric for the structure in question.

A variant of this approach is to use relative similarities in composition between different tissues as an indicator of possible common tissue origins. An example is that of the analysis by *Wegman* and *Gilman* (1990) of two blood cell tissues in comparison to the skin of the mouse. Mixing two input genotypes that differed for a white blood cell marker (an immunoglobulin), a red blood cell marker (hemoglobin), and a coat colour marker, they observed a correlation in nine chimeras between the composition of the immunoglobulin and hemoglobin markers, with a much smaller degree of correlation between these blood cell markers and that of coat colour. The correlation between white and red blood cell genotypes was inferred to indicate a common cellular origin of the white and red blood cell lineages. From the observed frequency of chimerism in each of the markers, primordial cell numbers of both hemopoietic and surface ectoderm were calculated as being no fewer than five cells each.

There are two principal difficulties with all such statistical estimates. The first is the most important; the calculations are based on the implicit assumption that there is *complete and random cell mixing until the moment of tissue foundation*. The consequences of this assumption were first described *by McLaren* (1992). If there is a period or regional coherent clonal growth preceding the moment of tissue allocation, then the estimate of founder cell number will in reality be an estimate of the number of marked coherent dones from which the primordium Originates. I he greater the: extent of coherent clonal growth receding tissue allocation, the lower the numerical estimate of cells (= clones) will be. Indeed, all such estimates for fetal tissues probably reduce to the number of *original fetal progenitor cells present of the introduction of genetic heterogeneity*. For chimeras made by morula aggregation, the estimates of presumptive founder cell number for a given fetal structure therefore, in all probability, refer to the number of cells present at aggregation that contribute descendants to the structure in question, or about five to eight cells.

The estimate of five to eight cells derives from the following considerations. From single cell injections into blastocysts, it is apparent the single ICM cells can contribute to most or all fetal tissues and hence organs. Therefore, most or all 1CM cells in the late blastocyst probably contribute to all fetal tissues. (Such widespread decendancy of single ICM cells presumably reflects a thorough mixing of cells prior to or during primitive streak formation.) Tracking prelabeled cells in morula aggregates, however, shows little cell mixing until blastocyst formation with the consequence that the ICM derives primarily from the five to eight cells that are present internally in the aggregate. Indeed, most binomially derived estimates of fetal tissues founder cell number are in the range of two to five or slightly more cells.

There is a second problem with the statistical approach: the influence of selection during cell growth within chimeras. When the cells of the two input strains differ in background genotype, they often contribute differentially to different tissues or organs. A consistent underrepresentation of one genotype in a tissue relative to its frequency in other tissues is a certain indication of relative tissue-specific differences in growth or competitive ability between cells of the two genotypes. Such selective differences can, of course, influence the relative contributions to several tissues, undermining the presumptive significance of correlative frequencies for inferences about similar or disjoint cellular origins.

The analysis of chimeric tissue in situ, using direct cell markers, eliminates some of the ambiguities inherent in the statistical approaches. By permitting the direct visualization of marked cells within the analyzed tissues, the *in situ* screening techniques directly yield information on clonal growth patterns. This information, in turn, can help of furnish some minimal estimates of founder cell numbers for the structures or tissues in question. When the screening is carried out only on adult or newborn mice, the inferences remain indirect and are affected by the uncertainty as to whether individual founder cells or clones or adjacent founder cells are being estimated. However, if the marker or markers of choice are expressed throughout development, and if screening is carried out throughout this period, the method can yield clear results. In practice, few such studies have yet been carried out. But the use of cell autonomous antigenic markers may make this approach feasible in the near future.

The first example of an *in situ* analysis of chimeric patterns concerned the origins of the melanocytes, the cells responsible for pigmentation of the mouse coal. The melanocytes are the daughter cells of the migratory mesen-chymal cells termed "melanoblasts" that originate in the neural crest, the region of epithelium that marks the site of dorsal closure of the neural tube. The neural crest gives rise to two major populations of migratory cells, those that migrate deep into the body between the neural tube and somites to give rise to neurouns of the autonomic nervous system, and a second population that migrates just underneath the surface of the ectoderm. It is the latter group that includes the melanoblasts.

By making chimeras between strains that, differ in melanocyte pigment-forming capacities, it is possible to. determine something about the clonal histories of the melanoblast cell populations. Such experiments were first performed by *Mintz* (1997). Chimeras were made between albino (c/c) and wild-type (C/C) strains and between strains differing in melanocyte genotype for colour (e.g., black *B/B* and brown b/b). In all of the experiments, the striking outcome was the production of striped chimeric mice. Initially, *Mintz* reported that the patterns showed a regular alternation of the different colours, but larger samples have shows that there can be considerable variability in the width of the stripes.

Despite this variability, an underlying or "archetypal" striping pattern can be descerned when all the chimeras are compared. The wider bands observed in the animals simply consist of multiple unit widths, where the unit width corresponds to the narrowest single stripe seen. The striping pattern itself is explicable if colouration in the coat is produced by small numbers of melanoblast clones arranged in an initial linear series along the longitudinal axis in which the clones that neighbour any particular clone may either be of the same genotype (giving a broader band) or of the alternative genotype (giving a central band of unit width). Comparison of the banding patterns in many chimeric mice suggests that the head region consists of three unit stripe widths, the body six, and the tail eight. Because the left and right sides at any position along the longitudinal axis can be of different colour, the clonal origins of the left and right sides must be independent. Equating stripe widths with founder clones, one obtains 17 founder clones per side, or a total of 34 melanoblast founder clones. The independence of the left and right sides suggests that the cells that give rise to melanocytes are committed to do so before dorsal closure of the neural tube is complete and also that the founder cells/clones begin their lateral migration before closure takes place. Were it otherwise, cross-contamination would produce a high frequency of mixed unit stripes. With dorsal closure beginning at 8.5-9.0 days p.c., the migration of melanoblasts must take place before this period.

Does the estimate of 34 melanoblast founder clones mean that there are only 34 founder cells of the melanoblast lineages? Not necessarily. From the fact that the clarity of the striping is sharper in chimeras composed of cells from more distantly related strains than from more closely related ones, it appears that there is a partial sorting out according to genotype. If a certain amount of self-segregation is occurring early in the melanoblast lineage, the founder clones might consist of two or more cells of like genotype. However, the number is unlikely to be more than two or three, since the higher the number the more likely would be the occurrence of "split" unit stripes, and these are not seen.

The technique of chimera analysis has also been used to explore the origins of mesodermal structures derived from the somites. It has been possible to analyze the clonal origins of vertebrae, for instance, because certain common laboratory strains differ in their details of vertebral morphology Utilizing these morphological differences as indices of genotype and making chimeric mice from such strains, *Moore* and *Mintz* (1992) concluded that both the left and right halves and the anterior and posterior halves of each vertebra have separate and distinct clonal origins; these findings indicate that each vertebra is derived from a minimum of four clones. This result fits the known facts of somite development well; each vertebra is formed by the fusion of the caudal (posterior) and cranial (anterior) halves of neighboring somites and by those of the left and right somite pairs. The estimate of four clones is a minimum estimate of the number of founder cells because of the possibilities of cell type self-assortment and differential selection within founding cell groups.

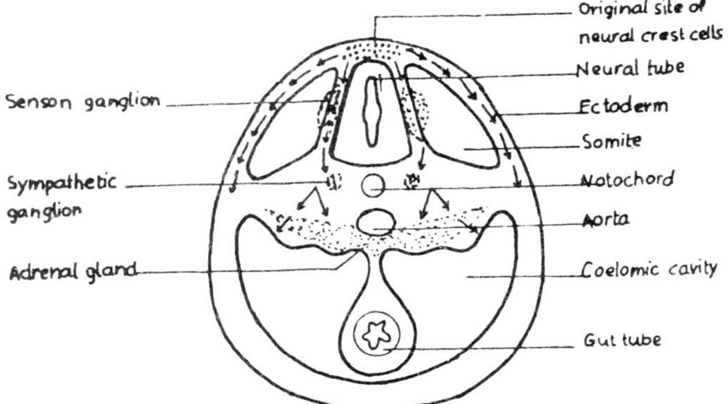

Figure 9.10 : Two pathways of neural crest migration. The outer arrows indicate cells that give rise to melanocytes; the internal arrows denote neural crest cells that give rise to parts of the neural system and the adrena! gland; stippled areas indicate neural crest-derived areas.

A different approach, using the indirect cell marker Glucose Phosphate Isomerase (GPI), was used to examine the origins of skeletal muscle: do these muscles arise by cell fusion or by repeated nuclear division without cytokinesis? The approach relies on the fact that many enzymes are made up of two or more polypeptide subunits. For such polymeric enzymes, the existence of two alleles specifying different electrophoretic variants within the same cell results in the production of "hybrid" enzyme molecules; these molecules are distinguishable from the parental enzyme forms by their intermediate position on gels after electrophoretic separation. If skeletal muscles arise by fusion of cells, then chimeric muscles from strains that differ in GPI alloenzymes should make hybrid GPI. On the other hand, if each syncytial muscle arises by repeated nuclear divisions from a single cell, then the muscles of chimeric mice should make only the two pure forms of the enzyme. The analysis of skeletal muscle from such chimeras revealed the presence of substantial amounts of hybrid enzyme in addition to the two parental alloenzyme forms. Muscle syncytia must. therefore arise through fusion of cells rather than as clones of individual nuclei.

In a later *study, Gearhart* and *Mintz* (1992) analyzed the origins of both somites and muscles, using GPI as the marker. They found that the great majority of individual somites from 8-to 9-day embryos contained both forms of GPI. Individual somites are therefore note clones, but originate from at least two mesodermal, cells. In the analysis of certain individual eye muscles, each of which was formed from a single myotome of a single somite, 22 out of 33 muscles assayed were found to contain both GPI alloenzymes and hybrid enzyme. The results show that each somite-derived myotome is also formed from two or more myoblasts. It was also observed that neighbouring somites may share some cellular ancestry.

All of the results discussed above have involved restrospective analyses of events: inferences about the earliest events are made from the tissue composition of the adults. Relatively few studies have attempted to determine the history of a tissue by tracing the development of marked clones in chimeric tissue. One example of such an analysis is that of *West* (1996) on the growth patterns of cells in the retinal pigment epithelium (RPE) of the eye. This monolyer of epithelial cells lies between the neural retina and the enclosing outer structure of the eyeball, the choroid. By making wild-type/albino chimeras and examining cross sections of the RPE, West showed that cell mixing in the RPE is extensive until about 12 days p.c. At that point, each clone is little more than one cell on average. This period of extensive cell mingling is then succeeded by a period of more coherent cell growth, with some cell migration taking place during continued growth of the monolayer. Final clone sizes of marked tissue seen in section consist of five to six cells.

Such one-dimensional sections through a monolayer give a final-gained picture of clonal growth. Examination of the total area of the RPE in chimeras provides a more complete picture of its history. When the two-dimensional surface of the PRE is analyzed in wild-type/albino chimeras in which the pigmented cell population in the RPE is a. minority component, distinct radial patterns of growth can be discerned. From the size of the sectors, and on the assumption that each sector corresponds to a founding clone, the results indicated that the RPE is initiated by a relatively small number of founding clones, probably more than 10 but fewer than 250. Synthesizing the cross-sectional and whole surface observations, it appears that the RPE is founded by a relatively small group of founder cells that generate descendants in a radial pattern of outward growth, initially with much local cell mixing the subsequently with greater coherent clonal growth.

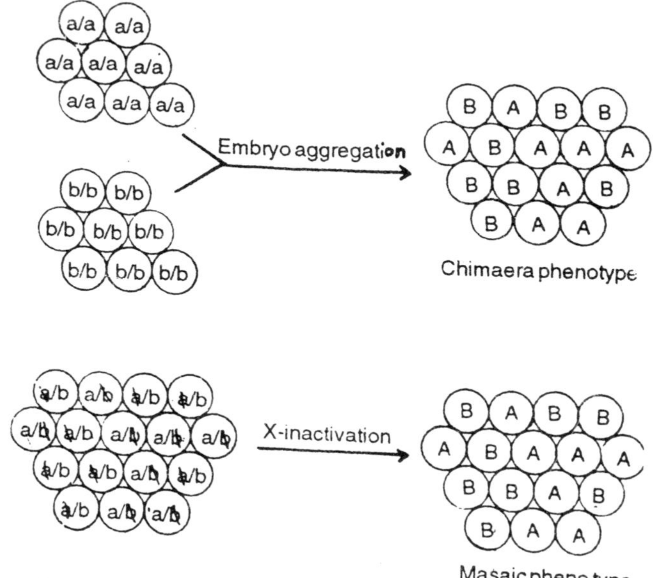

Figure 9.11 : Two ways to produce mice with a dual-composition cell population. Top, chimera construction bottom. X chromosome inactivation in an X chromosome heterozygote.

X Chromosome Inactivation Analyses

The naturally occurring phenomenon of X chromosome

inactivation in female mammalian embryos furnishes the second major form of clonal analysis in the mouse. Although X inactivation, like chimera production produces two phenotypically different populations of cells, the difference is generated intracellularly within individual heterozygous embryos rather than by the forced mixture of cells from two genegically distinct embryos at the outset. An additional difference between the methods is that X inactivation analyses usually require the use of X chromosome markers, while chimera analysis can employ any chromosomal markers, whether autosomal or sex-linked. The few autosomal genes that can be utilized in X chromosome mosaicism studies are those located on pieces of autosomes that have been translocated to the X.

These translocated fragments experience the X-inactivation process, and if the translocated piece and the normal homologue differ for a scorable marker, a mosaic pattern of expression is produced. The most useful of these translocations is Cattanach's translocation, which is a transposition of part of chromosome 7, containing the wild-type allele for the albino locus (C), into the X. When the standard seventh chromosome homologue contains the recessive (C) allele, inactivation of the translocated chromosome produces an albino clone in the descendants of that cell.

As drawn schematically, the phenotypic outcomes of chimera construction and X inactivation look identical. In reality, the patterns of spatial heterogeneity in chimeras and X inactivation mosaics often look somewhat different even when the same markers are employed. In general, the chimeras show more variable patchiness and larger patches. Indeed, the two properties are directly related. Variable tissue composition between cells of two genotypes tends to produce larger patches, with any shift from a 1: 1 ratio increasing the probability that cells of like genotype will find themselves as neighbours. However, when cell mixing is extensive, as in the early period of development of the RPE, the differences between chimeras and X chromosome mosaics are reduced. Only when there are large differences in background genotype between the component strains of chimeras does the chimeric pattern begin to look different from the X mosaicism pattern because of selective self-assortment.

The principal methods for estimating founder cell number in X chromosome mosaics are statistical. The central premise of these methods is that if sampling for tissue foundation is from a randomly mixed cell pool, the variance in composition for a given tissue between different mosaics in the same series will be an inverse function of founder cell number. When the founder cell number (n) is very small, a large variance between like samples will be observed; when *n is* large, the variance should be small. If tissues or organs show similar compositions and similar variances, they may stem from a common precursor pool; the residual variance between different samples would then reflect differences in the founder cell numbers of the individual primordia.

The first study of this kind was reported *by Nesbitt in* 1971. The cell marker she employed was Cattanach's translocation, whose frequency of inactivation can be scored in chromosome spreads from mitotic cells labeled with [^{3}H]thymidine early in S phase and scored autoradiographically. (Because the inactive, heterochromatinized X replicates late in S phase. The portion of the cell cycle devoted to chromosome replication, it appears as the nonlabeled X in these conditions). Samples were prepared from several different tissues-spleen, lung, liver, and coat melanocytes-of newborn mice, cultured for 4-5 days to amplify the fraction of dividing cells, and then analyzed by the mitotic spread procedure. The analysis indicated that there are about 20 epiblast cells that act as the precursor of the fetus at the time of X chromosome inactivation and that 20-50 founder cells are allocated within the primitive streak for each of the organ systems analyzed.

A more recent approach to the same problem has been reported by *McMahon et al.* (1983). They used an indirect cell marker, the X-linked enzyme Phospho Glycerate Kinase (PGK), which exists in, two electrophoretically distinguishable forms, and a sensitive assay, suitable for small amounts of electrophoresed enzyme from tissue samples were prepared from 12.5-day female embryos, heterozygous for the two alleles of PGK, and analyzed for PGK electromorph composition. The tissues examined were representatives of. the three primary germ layers of the fetus-neural ectoderm, heart mesoderm, and liver endoderm-and the germ cells. The analysis indicated that there are 47 fetal precursor, cells and about 190 precursor cells for each of the four tissues examined (at the time of germ layer delimitation within the primitive streak). Although the last number is an average, a high correlation of alloenzyme composition between different tissues from each embryo was observed suggesting that all three primary germ layers and the germ cells arise from similar-sized precursor cell pools. The higher estimates of founder cell number by *McMahon et al.* (1983), compared to those of *Nesbitt,* (1971), reflect the lower variances observed in their study.

The estimate of 190 germs line precursor cells (PGCs) is of special interest for two reasons. Firstly, this number is considerably greater than the number of PGCs detectable by AP staining at *8.5* days, the first point at which the cells can be observed directly. If the statistical estimate is correct, then high AP activity may identify only a fraction of the true number of PGCs. Secondly, the similarity of this number to that for the other three germ layers raises the possibility that PGCs are not actively instructed to be such but may simply be the residuum of unassigned cells. In this view, the retention of totipotency by PGCs is a consequence of their failure to be restricted in fate. It may even be that the caudal end of the primitive streak, from which the PGCs arise, is a region that protects against restriction of totipotency. As discussed earlier in connection with germ line determination in *Drosophila,* the posterior polar plasm from which the germ line of the fruit fly originates may exert its determinant function by providing comparable protection.

As in the chimera studies, the estimates of founder cell number from X mosaicism studies are probably estimates of the number of coherent clones contributing to the tissue rather than the number of founder cells per se. When more reliable information from other techniques on the times of tissue allocation becomes available, both forms of clonal analysis should contribute more precise estimates of founder cell number than are presently obtainable.

DEFECTS IN ORGANOGENESIS: GENETIC ANALYSIS

A very large number of mouse mutants are affected in organogenesis Many of these mutants were first detected on the basis of their dominant heterozyogous phenotypes-for instance, a degree of runting or a change in coat colour from the parental stock. In homozygotes, the mutant defects often result in embryonic or perinatal mortality. Another class of mutants affected in the development of organ systems are those initially identified on the basis of their behavioural phenotypes; such mutants are usually defective in the nervous system. The *shaker (sh)* and *waltzer (w)* mutants, whose phenotypes are conveyed by their names, are examples of neural defectives.

Until the advent of chimera and mosaic studies, the developmental genetics of the mouse consisted almost entirely of the study of the developmental defects associated with particular mutants. The procedure is the classical one of developmental genetics. One first identifies the set of mutant phenes and then works backward in development to the first observable aberrancy. From these observations, one then attempts to reconstruct the cellular site, and if possible the biochemical basis, of the initial developmental derangement. In a number of instances, however, the etiology is similar to that produced by known specific enzymatic deficiencies in humans and the mouse syndrome proves to have an identical basis.

For the most part, however, the mutant defects that produce severe aberrations of early or middevelopment have not been characterized at the biochemical level. Nevertheless, these mutants represent a valuable store of genetic source material for future investigations. As the techniques for biochemical and cytological characterization advance, the precise biochemical basis of some of these mutant defects will become clear and the characterization of the genes themselves, isolated via recombinant DNA procedures, will elucidate the genetic basis of individual defects in an increasing number of instances.

Perhaps the single most valuable tool in characterizing the cellular basis of mutant defects has been the electron microscope. With electron micrographs, it is often feasible to determine which cells are defective and to reduce the range of possible explanations of the mutant phenotype (although ambiguities usually still remain). An example of such an analysis will illustrate the usefulness of the approach.

One of the first morphological mutants of the mouse to be characterized was the dominant short-tailed *T* or *Brachyury* mutant, reported by N. *Dobrovolskaia-Zavadskai in* 1957; When short-tailed (T/+) mice are crossed. one obtains a 2: 1 ratio of short tailed to normal mice instead of the expected 3:1 dominant to recessive ratio (1T/T :2T/:1 +/+).This observation suggested that, like A. T is lethal when homozygous. This suggestion was confirmed *by Chesley* (1985), who performed the first detailed examination of the embryos. Using the standard light microscopic techniques available at the time, *Chesley* observed that the first abnormalities in *TIT* embryos appeared in early somite, 8.5 day p.c. embryos. In the posterior third of the embryo, the notochord is absent, the somites are disorganized, and the neural tube shows aberrant morphology. These early defects are succeeded by progressive disorganization of the entire notochordal and neural tube regions and a failure of hindlimb development. Heterozygous (T/+) litter mates, incontrast, do not show any defects until about 11 days p.c. At this point, they exhibit a diminished neural tube and notochord in the tail region and minor irregularities of the notochord anteriority. The heterozygous syndrome is thus a milder version, with a later onset, of the homozygous developmental defect. The cause-effect relationships between the various abnormalities could not be determined, but Chesley speculated that the notochordal defect was primary and the neural tube defects secondary.

In a detailed electron microscopic study, *Spiegelman (1996)* confirmed and extended the earlier findings. By 8 days p.c., the T/T embryo has a reduced trunk and a correspondingly enlarged primitive streak. In these embryos, the conversion of embryonic ectoderm to mesoderm is blocked; in consequence, instead of forming notochord and somites in the posterior region, the blocked mesodermal cells form excess neural tube material. In effect, the block in mesodermal differentiation from ectoderm leads to a temporary hyperplasia or overgrowth of neural tissue: these epithelial tubes subsequently undergo degeneration. concomitantly with the disintegration of the anterior mesodermal axial structures. (This syndrome of neural overgrowth.

The specific cellar defect producing the T-lethal syndrome is revealed by the electron microscope. Unlike the wild-type embryo, in which the neural tube and notochord are separated by a discrete basal lamina, these two cell types meet in the lethal T embryos at the ventral margin of the neural tube. There are nobasal laminae in the region of contact, and the neural tube and notochordal cells establish junctions with one another. This failure of normal cells recognition, with subsequent failure to establish a normal barrier between the notochord and the neural tube, must play a role in the ensuing multiple failures of mesodermal organization and differentiation.

The difference in notochordal cell surface behaviour in the mutant could be a direct reflection of a cell surface property that directly impedes normal primitive streak differentiation. It is equally possible that the cell surface change is a secondary event and perhaps only one element of the failure of mesodermal differentiation. The pictures themselves do not settle this point, but only raise it.

To characterize the role of a wild-type gene in development, one requires a complete description of both the cellular and developmental phenotype and the genetic character of the 'nutation itself. With *Drosophila,* one can usually perform the Mullerian dosage test in order to classify the expression property of the mutation in the absence of the necessary duplications and deficiencies, one can resort to isolating additional mutant alleles of of same gene. The various mutants can then be compared, and the phenotype produced by a complete or nearly complete inactivation of the gene can be inferred.

In the mouse, these genetic stratagems are usually impossible, for the most part, the necessary duplications and deficiencies are not available for the dosage test, and largescale searches for non-complementing lethal mutants are impractical.

Only when a mutation creates a distinctive visible phenotype without attendant lethality can one hope to isolate more mutants of the same gene by direct screening, permitting comparative studies. To isolate new mutants, one screens for non-complernenter chromosomes, as in *Drosophila.* Heterozygotes carrying the initial mutation (+/m) are mated to mutagenized wild-type animals, and the progeny are examined. Any individual in the F, showing the mutant phenotype must carry the original mutation m, and a new mutant allele, m". These individuals are then outcrossed to isolate the new mutant, which can be distinguished from the original if the original mutation is linked to scorable markers.

Because this method is limited to genes in the mouse that give a visible but fully viable phenotype, it is inapplicable to many genes of developmental interest, Where it can be applied, it has been informative. The procédure is illustrated by the search for mutants of the *albino (c) locus* on chromosome 7. This gene is probably the structural gene for tyrosinase, the enzyme that synthesizes the major melanic,pigments. To collect new albino mutants, female heterozygotes (c/C). The C allele being the dominant, wild-type allele) were mated with irradiated wild-type males, and all albino progeny were outcrosed to isolate the new c alleles. A number of these new mutants are lethal when homozygous. The reason for the lethality is not the absence of c activity but the delection of the mutant chromosomes for adjacent essential loci.

Gluecksohn-Waelsch (1999) has described a set of six of these lethal deficiency *albino* chromosomes. One mutant homozygote dies in early cleavage, one dies in the early egg cylinder stage, and the remaining four exhibit perinatal mortality; these last four exhibit a specific deficiency for six particular liver enzyme activities (while having normal activities for several other liver-specific enzymes). By making heterozygotes of pairs of the different lethal deficiency chromosomes and scoring for the presence or absence of particular mutant phenotypes, one can position the wild-type genes whose activity prevents particular mutant phenes. Such a gene localization procedure is termed "Complementation mapping." The liver enzyme defect is between c and *Mod-2* (the gene for mitochondrial malic enzyme-2), and the essential early embryonic function(s) lie between *Mod-2* and sh-1 (one of the genes that gives the shaker phenotype) the results show that isolating new mutants of a locus can allow one, if some of the mutations are deletions, to work "outward" along the chromosome from that locus, identifying new essential gene functions and their linear sequence. In effect, the locus provides a bridgehead for exploring the genetic organization of a chromosomal region. (An analogous molecular approach to the analysis of chromosome genetic organization.

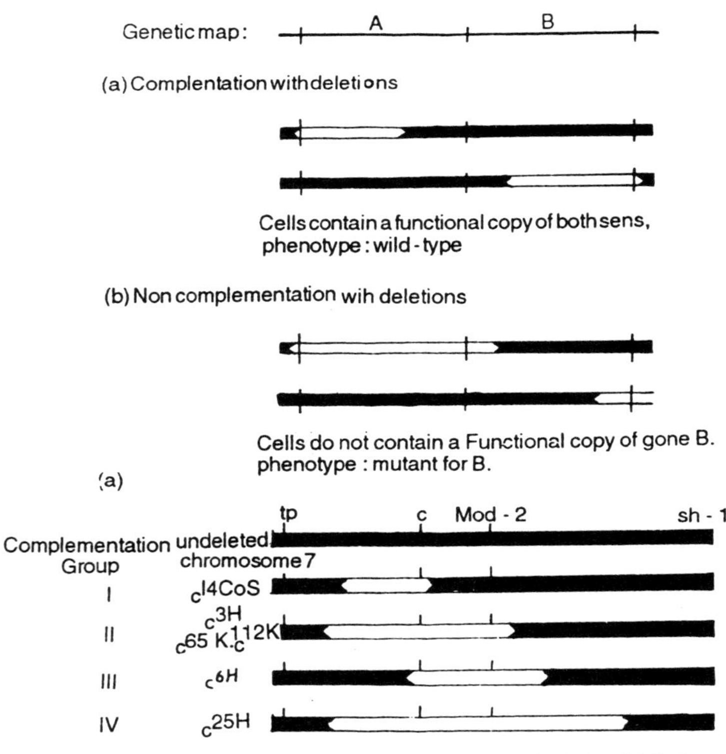

Figure 9.12 : Delection mapping of functions (a) principle of deletion mapping; (b) deletion map of c region.

The defect in the four perinatal mutants may be of special interest with respect to genetic regulation. In these mutants, the cause of death appears to be a combination of defects in the liver, kidney, and thymus, with the liver defects being the most serious. The enzymatic deficiencies of the mutants may reflect the loss of a specific regulatory locus essential for their activity. When mouse/rat hybrid cells are constructed by fusing mouse cells homozygous for one of the deletions with rat cells, the hybrid cells are found to express the mouse G6PD activity (one of the deficient activities of the mouse cell); evidently, the rat genome can supply the missing regulatory factor. The immediate cause of the inability to synthesize the affected enzyme activities may be a reduction in receptors for glucocorticoid hormones found in the mutant cells. However, the presence of some receptors in the cells of the deletion homozygotes suggests that the structural gene for the receptors is not missing.

The mortality effects associated with the late-acting mutants may involve certain aberrancies of cellular organelle development. Electron microscopic inspection of the affected organs in the four perinatal lethal homozygotes shows that the membranes of the rough endoplasmic reticulum, the Golgi apparatus, and the nucleus are all abnormal in the liver and the kidney. However, these membrane organelles are normal in appearance in the thymus of the affected embryos, although the thymus itself shows morphological abnormalities. These organelle effects reveal tissue-specific differentiative aspects of certain commonly shared organelles. The existence of such subtle differences had not been anticipated on the basis of conventional cytological studies and adds a further dimension to the known complexity of tissue differentiation. However, some caution is required in accepting this interpretation: the liver organelle defects may be particular secondary consequences of the enzyme or other deficiencies that do not pertain to the thymus.

THE *T*-COMPLEX: MAMMALIAN SWITCH GENES OR DEVELOPMENTAL ANOMALY?

One of the fundamental questions in animal developmental genetics is whether development is governed by major regulatory "switch" genes. We have seen instances of apparent switch genes in the two invertebrate systems discussed earlier. In the nematode, replacement of various cellular lineages can be produced by mutation in a few specific genes, with consequent major alterations in the developmental pathway. And in Drosophila, the homeotic mutants, particularly those of the bithorax and *Antennapedia* gene complexes, show that certain genes play key roles in the assignment of segmental phenotype. Are there comparable loci in mammalian development?

During the 1970s, several investigators proposed that the so-called t-complex on chromosome 17 of the mouse played just such a key regulatory role in early mouse development. Several results have rendered this hypothesis increasingly unlikely, but the developmental genetics of the t-complex have produced much information of interest.

The first *t* mutations bearing chromosomes were picked up fortuitously because of an unusual reaction they display with *T*. As described earlier, *T/+* heterozygotes are shorttailed mice and *TIT* homozygotes are embryonic; lethals. When certain mice from wild populations were crossed with T, a new class of viable progeny appeared: completely trailless mice. Surprisingly, the condition of taillessness was found to breed true in brother-sister matings from each cross; that is, only tailless progeny were produced in each ensuing generation. Outcrossing revealed that each tailless mouse carried a new recessive lethal, a *t* mutant, in addition to T Crude mapping placed the new mutants near T on the chromosome now designated number 17. Each tailless strain therefore carries two heterozygous lethals. *t* and *T*. Such strains, in which each member of a homologous chromosome pair carries a lethal mutation and in which only the heterozygous combination is viable, is said to be a balanced lethal strain. The explanation for the perpetuation of the tailless phenotype is that the T/T and t/t embryos are lethal and only tailless heterozygotes (T/t)survive.

The genetic complexity of the T/t locus, as it was first designated, emerged only when the different tailless strains were crossed with one another. If all of the t mutations are in the same functional genetic unit, then the intercrossess should yield only tailless mice. Thus, if two different t mutants are designated *t'* and *tb*, the intercross *(T/t^b)* should give only parental genotype, tailless progeny. In fact, some crosses gave normal-tailed mice in as many as one-third of the surviving progeny. These mice were subsequ-ently shown to be of genotype t^a/t^b. In other words, some of the *t* mutants were found to complement each other, permitting normal development. As more *t* mutants were isolated, it became apparent that the different lethal alleles could be placed into six different complementation groups.

In addition to t mutants from wild populations, new *t* mutants were isolated in the laboratory from the true-breeding tailless T/t) lines, at a rate of about 1 in every 500 or 1000 progeny. These new mutants were isolated as exceptional, normal tailed mice within the stocks, and were shown by the complementation test to carry new *t* alleles, derived by some form of unequal crossing over within the *T/t* complex. Many of these laboratory-derived strains are semi- and fully viable *t* mutants, each of the new mutants being identified as a *t* mutant by the T interaction test.

However, direct genetic mapping of the mutant sites could not be carried out because all of the *t* mutants exhibited a region of depressed recombination within the area of the tcomplex itself. Furthermore, many lethal t mutants show unusual inheritance patterns, in particular, a tendency for the mutant allele to be preferentially transmitted from wild-type heterozygotes (t/+), but only in the male line. In contrast to this behaviour of the lethals, some of the viable t/'s are present in less than 50% of the sperm of heterozygotes. An additional complexity is that combinations of high- and low-transmission complementing t alleles do not always behave predictably, but can show partially reversed relative transmissibility in each other's presence.

The causes of the these distortion in transmission ratio are obscure but reflect both early and late developmental effects on spermatogenesis. The perferential transmission effects from t/+ heterozygotes apparently reflect direct alterations in the fertilizing ability of the haploid mature sperm. These changes in sperm behaviour are correlated with specific antigenic changes in the sperm head. Each lethal *complementation t group* is associated with a particular set of sperm antigenic determinants; some antigenic determinants are shared between different groups, while other are unique to a particular complementation group.

The potential developmental significance of the t-complex

became apparent only when the specific lethal syndromes of homozygous inviable embryos were closely examined. It was found that members of each complementation group show arrest of embryonic development at characteristic stages, ranging from early morula to early somitogenesis. The developmental blocks associated with each lethal complementation group, different mutants often show some differences in severity or terminal stage, but the members of each group resemble each other more closely than do members of different groups.

TABLE 9.2 : DEVELOPMENTAL BLOCKS IN LETHAL *t* MUTANTS

tComplementation Group	*Stage of First Abnormality*	*Stage of Developmental Arrest*	*Cell types Affected*
$t^{1/2}$	Morula	Morula	All blastomeres
t^{w73}	Late blastocyst	Early egg	TE Cylinder
t^{o}	Early egg cylinder	Late egg cylinder	TE. EmE, endoderm
t^{w5}	Late egg cylinder	Preprimitive streak	Mainly EmE
t^{9}	Early primitive streak	Late primitive streak	Mesoderm and EmE
t^{w1}	Late Primitive streak	Early organ genesis	Ventral neural tube, brain

In a review of the genetics and biology of the t mutants. *Bennett* (1995) proposed that the developmental effects reflect a major governing role of the presumed wild-type t-complex in early mouse development. This hypothesis amplified an earlier suggestion made by *Glucksohn- Waelsch* and *Erickson* (1990). Based on the known effects of the t mutants on sperm antigens *Bennett* proposed that normal development involves a fixed sequence of changing cell surface properties, each essential for the next developmental transition. The role of the t-complex would be to determine this programmed sequence of changing cell surface properties.

Because the earlier embryological studies pinpointed ectodermal cells and derivatives as the chief foci fort mutant effects, the action of the t-complex was provisionally characterized as the mediation of a series of switches within the ectodermal lineages. In terms of this hypothesis, the developmental effects of T were interpreted as stemming from the cell surface defect observed in the electron micrographs.

The most telling argument against the developmental switch hypothesis comes from an analysis of the expression oft antigens during development. If the developmental transitions in the early embryo are mediated by the appearance of specific cell surface molecules on particular cell types at specific developmental stages, then one would predict that the sequence oft mutant syndromes, from morula to late neural tube stages, would be reflected in a sequence of t antigen appearances on the cell surface(s). *Kemler et al.* (1996) tested this hypothesis by preparing antisera specific to the various t complementation groups and monitoring the time course of appearance of the t antigens on embryos. (The antisera were prepared by immunizing mice with spermatozoa from heterozygotes of the various t lethals and then absorbing the antisera first with wild type sperm and then with lymphocytes, the latter step to remove other non-specific antigenic determinants.) The time course of appearance of the different antigens on heterozygous (t/+) embryos was then monitored by treating the embryos first with the specific sera and then with a fluorescent antimouse Ig antibody. Under these conditions, only embryos, with the characteristic t antigen on their surface fluoresce.

For four mutants representing four different t lethal complementation groups (the t^{12}, t^{0}, t^{w5} and t^{9} groups), t-specific antigens were detected on embryos as early as the eight-cell stage in all cases. By this test, the t mutant cell surface changes all appear early and without temporal relationship to the time of onset of the various developmental aberrancies.

The developmental switch hypothesis also requires the existence of wild-type alleles of the *t "mutations."* Much evidence now indicates that the t mutants differ from the wild-type in large blocks of DNA within the proximal region of chromosome 17, rather than in single-site mutations. The absence of recombination between wild-type chromosomes 17 and t mutant chromosomes in this region reflects their absence of finestructure homology. Instead of mutant alleles, therefore, particular *t* mutations are said to be *"t haplotypes."* Furthermore, the different developmental aspects of *the t* mutants phenotype-interaction with the T locus, embryonic lethality effects, enhanced transmission ratio in T/+ heterozygotes, male sterility in many double heterozygotes—are caused by different loci within the block of t-chromatin. It has been possible to dissect these different loci genetically because of the occurrence of rare crossovers in the promimal region between t haplotypes and marked wild-type chromosomes; the recombinant chromosomes bearing portions of the t- complex experience recombination with other t chromosomes at normal rates, because of their shared oenology, and can be further analyzed. A genetic map of the t region, and its wild-type analogue has been constructed by *Siler* (1981). The results explain why no t mutants have ever been isolated from non- t carrying populations in the laboratory. In contrast, several x-ray-induced mutants of T, which was once thought to be part of the t-complex, are known.

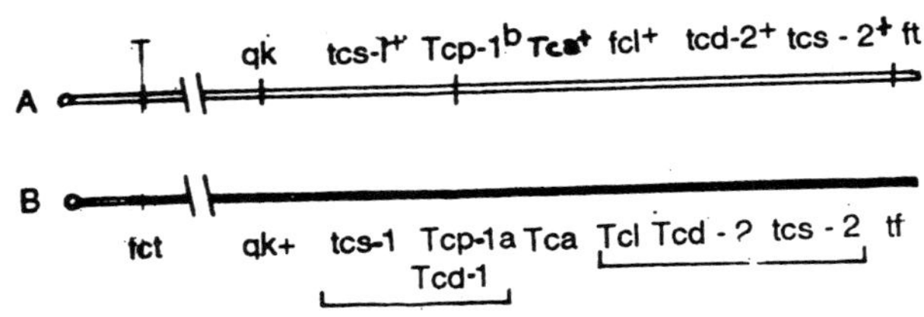

Figure 9.13. Map of the t-complex region in wild-type and complete t haplotype. The solid line represents t chromatin. T, shorttailed : tct, tail interaction factor, qk, quaking; tes-1 and tsc-2 proximal and distal sterility factors, respectively ; Top-1. testicular t complex protein:* Tcd *and* Tcd-2, *proximal and distal segregation distortion factors respectively; tcl., t-lethal factor(s) ; If. tufted.

The developmental differences between *t* haplotypes reflect differences in the amount and sequence composition of t chromatin they contain. Indeed, the observations suggest an unusual origin oft chromosomal material, originally proposed by *Dunn* and *Bennett* in 1991, namely, that all *t* mutants share a common ancestral chromosome introduced from a distinct and partially diverged mouse population. Because *t* mutants are found in all wild *Mus musculus* populations, the hypothesis demands that the progenitor chromosome has spread throughout the world via these popula-tions. Such diffusion is explicable in terms of the genetic properties of the complex itself; recombination suppression would tend to keep the transmission-enhancing factors together with the lethality and sterility factors, with the former acting to keep the *t* chromos-omes at high frequency within each population. Only the delete-rious effects associated with the *t* chromosome out of the genetic pool.

The direct evidence that all *t* chromosomes derive from a single source is a major testicular protein encoded by a gene *(tcp-1)* in the region of the *t* complex. All non-t-bearing mouse strains. including T/+ strains, produce a slightly basic form of the protein, while all t mutants possessing the proximal region of the complex produce an acidic form of this testicular protein.

The *t*-complex is therefore extremely interesting from the dual perspectives of population genetics and genome evolution. The basis of the developmental effects, however, remains elusive. The t-complex contains analogues of several genes within the comparable region of the wild-type chromosome. yet homozygosity for particular sets of these *t* gene produces products distinct developmental aberrancies. In some sense, the gene activities of the *t*-complex are mismatched with the developmental activities that arise from the rest of the genome of M. musculus, in effect a form of hybrid lethality. Such lethality is usually believed to reflect a failure of the transcriptional regulatory apparatus to cope with gene products or regulatory sites that are slightly different.

Many questions about the *t*-complex and *t* mutants remain. Solutions to the puzzle of how particular *t* lethals affect critical early transitions, in the mouse embryo may prove highly informative about these transitions, but the concept that the complex consists of a set of switch genes that operate in normal development is probably incorrect.

TERATOCARCINOMAS: MODEL SYSTEMS FOR DIFFERENTIATION

Some of the most interesting events in postimplantation development are those of tissue allocation and determination and the beginning of tissue differentiation. Unfortunately, these are also among the events most inaccessible to the experimenter. The methods of clonal analysis furnish some information about tissue foundation, but the conclusions are necessarily indirect and tentative and, by their very nature, cannot elucidate the underlying mechanisms in these early cell assignments. On the other hand, direct biochemical analyses of the first commitments and tissue differentiations in primitive streak embryos are rendered difficult or impossible because of the limited amounts of material available.

In recent years, an alternative approach to early development has been made possible by the realization that EC cells, a class of neoplastic stem cell, many provide an analogue of the early totipotent cells of the mouse embryo. EC cells were initially identified as the stem cells of certain tumorous growths called *"teratocarcinomas."* As the name implies, these growths are both tumorous and *"monstrous."* First discovered in humans, they appear to be highly disorganized embryos consisting of a nearly random mixture of tissues and structures such as teeth and hairs. The first class of teratocarcinomas discovered in mice was reported by L.C. *Stevens* and his colleagues in the 1950s. These were testicular tumors arising at a high frequency in the inbred mouse line 129. Analysis of these growth revealed that they certain two classes of cells, the various differentiated cell types and a core of cancerous stem cell, the EC cells. Culturing of the tumors showed that the differentiated cells are benign (non-cancerous), while the EC cells are neoplastic. In general, serial culture of teratocarcinomas in vivo leads to a loss of virulence expressed in the production of benign growths called *"teratomas."* However, the EC cells can be maintained as cell lines by serial culture in the intraperitoneal space. Such in vivo ascitic cultures maintain their malignancy and differentiative capacities indefinitely.

The first teratocarcinomas to be described were testicular tumors. However, EC cells can be derived from any mouse strain by transplantation of early (3- to 7-day) embryos to nonuterine sites, such as the kidney or testis capsule of the male adult. It is also possible now to derive EC lines directly from early embryos using in vitro culture. The frequency with which embryos give rise to teratocarcinomas is a function of their own genotype-different inbred lines giving different results-and of their maternal source, suggesting that there is a maternal effect in the potentiation to form EC cell. Furthermore, spontaneous ovarian teratocarcinomas occur with measurable frequency in one strain; these apparently arise from early oocytes. Teratocarcinomas can therefore develop from either male or female germ line cells, although always from diploid germ line precursors, or form early embryos. When the source is an early embryo, the neoplastic stem cells presumably originate either from the normal totipotent cells of the ICM or forth the primary EmE. Regardless of source, all EC lines can be cultured in vitro. Depending upon its history and a variety of little understood factors, each line comes to be characterized by a very broad, intermediate, or narrow range of differentiative capabilities.

When cultured intraperitoneally, EC cells typically give rise to two kinds of embryonic bodies. The First, simple embryoid bodies, consist of a core of EC cells surrounded by a layer of endoderm; superficially, they resemble the ball of ICM cells covered by endoderm, typical of the early blastocyst stage. The second class, cystic embryoid bodies, arise after more prolonged culture of EC cells and probably develop from simple embryoid bodies. They resemble the embryonic region of late egg cylinder embryos, but are substantially more disorganized and exhibit a mixture of differentiated tissue regions. Certain in vitro EC lines also differentiate in suspension, under the appropriate conditions, and produce both kinds of embryoid bodies. Other EC lines typically show differentiation in vitro in monolayers on the inner surface of the culture vessel.

Although EC cells resemble early embryonic cell and can be derived from them, there has been some debate as to which specific embryonic cell type they most resemble. From their patterns of synthesized polypeptides, they appear closer to primary EmE in developmental state than to ICM.

The most productive approaches to the characterization of early embryonic differentiation through EC cells have been immunological. In one procedure, rabbits are injected with EC cell s to produce an anti-EC serum, and the purified antiserum is then applied to staged early embryos or to other EC lines. The presence or absence of the target antigen on the embryos or cells being screened is determined either by the direction of a biological effect or by using fluorescein-conjugated reagents that recognize and bind to the first antibody. [The latter approach was used by *Kemler et al.* (1996) to detect *t* specific antigens.] A different strategy is to prepare antibodies to purified components found only incertain differentiated cells and then to monitor the time course of appearance of these antigens, usually by means of the fluorescence assay, in differentiating cells lines.

An example of the immunological strategy for identifying events in early development using EC cells involved the production of an antiserum to one nullipotential line of cells, cell line F9 the F9 antiserum has served to identify a cell component required for compaction of the early embryo. This antiserum is complex, but its specific gamma globulin component detects a cell surface molecule present-throughout early development. It is present on the surface of all cells of premorula embryos and in early and late blastocyst stages, becomes localized subsequently to the EmE, and finally disappears on the ninth day of development. In adults. The F9 antigen is found only on cells of the adult male germ line, including sperm cells. When early embryos are treated with the antigen-binding fragment of the gamma globulin fraction, compaction of the eight-cell embryo is prevented. Both the anti-F9 serum and several monoclonal antibodies that prevent compaction bind to a specific glycoprotein termed "uvomorulin." Uvomorulin undergoes a calcium ion-mediated conformational change that is essential for the process of compaction; the antibodies that block compaction evidently prevent this conformational change. Although the structural gene for uvomorulin has not been identified, two lethal *t* complementation groups block the appearance of F9 antigen. These *t* mutations presumably interfere with either the synthesis or surface localization of uvomorulin.

From the standpoint of genetics, the interest of EC cells lies in their potential use as vehicles for genetic alteration of mouse development. This possibility exists because the cells of certain EC lines can participate in normal embryonic development when injected into blastocysts, contributing descendant cells to all tissues and organs. This ability also shows that the neoplastic state of the EC cell is reversible when the cell is placed in a normal embryonic environment.

If normal EC cells can be reinserted into the developmental program, EC cells that are mutant but viable in culture may be similarly capable of reintegration. If so, the way is open for the construction of new lines of mice bearing mutations induced in vitro. The only conditions for success are that the introduced mutant cells be capable cf participating in somatic development and giving rise to germ line-tissue in the viable chimeras. This procedure, if successful, would be considerably more efficient than standard mutant screening of whole animals because vastly more cells can be examined in culture for mutant clones than whole animals screened for mutant pheno-types.

The first demonstration that EC cell can participate in whole animal development was reported by *Brinster* (1994). His experiment involved the injection of *agouti* cells into blastocysts of an albino strain. The source of EC cells that *Brinster* used was the ascitic fluid of a tumor line derived from strain *129,* serially propagated over many years by intraperitoneal injection. Of 60 surviving animals, each derived from a blastocyst injected with two to four EC cells, one animal showed a coat with partial gray striping. The source of this agouti colour must have been the donor EC cells.

This particular animal did not sire agouti progeny when mated with an albino female-the dominant *agouti* allele would have been detected in any progeny from agouti germ line cells-and was ,therefore probably not a germ line chimera. However, a germ line chimera was subsequently obtained by *Mintz* and *Illmensee* (1995), who used the same ascites tumor as donor. They obtained three chimeric animals, detected by patches of donor coat colour, which were chimeric in their red and white blood cell-forming tissues and in their livers (as shown by the presence of donor cell GPI. One of these mice, a male nicknamed *"Terry Tom,"* sired progeny bearing tumor cell line markers and was therefore a germ line chimera.

The 129 cell line is an in vivo line of EC cells, using EC cells to transmit new mutations requires the utilization of EC cells from individual cell clones, and such clones can be obtained only from *in vitro* EC cell lines Somewhat surprisingly, the great majority of such cell lines, when tested, including those with large pluripotential somatic developmental capacities and lacking any detectable chromosomal abnormalities, have failed to give germ lines chimeras. Either these cell lines are genetically deficient in one or more ways or the PGCs derived from them fail to migrate to the gonadal ridges in sufficient number.

However, successful transmission of an in-vitro cultured cell line MELT-1 (for "mouse euploid totipotent teratocarcinoma1") has been reported *by Mintz* and *Stewart* (1981). The cell line, also derived from strain 129, shows a normal karyotype and is chromosomally female; it was repeatedly passaged in vitro before being injected

into blastocysts of a host non-agouti (black) strain. Measured either in terms of the percentage of visible chimeras (as scored by the frequency of agouti contribution to the costs of the surviving animals) or by the extent of the contribution of donor genotype cells to internal tissues (as shown by the GPI marker), this cell line is significantly more normal than other previously tested in vitro lines. Of nine coat colour female chimeras, one proved to have strain 129 oocytes, giving rise to three agouti F_1 progeny. These three animals in turn gave rise to agouti F_2 progeny.

A similar approach with embryo-derived pluripotential (EK) cells may be even more efficient. Bradley et al. (1984) report that over 50% of surviving blastocysts injected with EK cells are visibly chimeric. Of these, approximately one-thud proved to be germ line chimeras. Induction of new mutations in EK cells for passage into the germ line promises to be superior to the use of EC cells in this respect.

It may soon be possible to transmit mutations, induced in EC

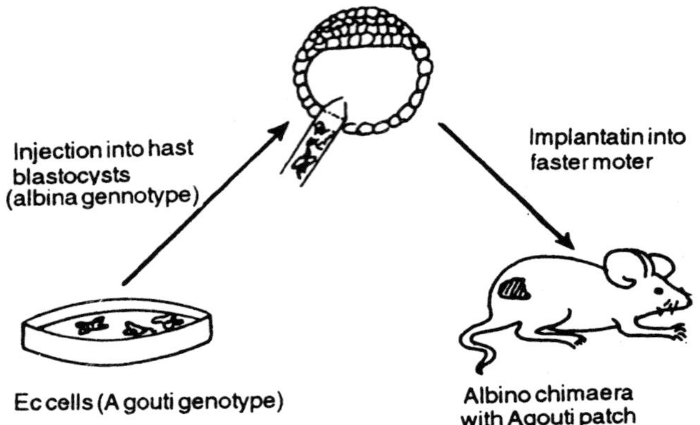

Figure 9.14 : Transmission of agouti-genotype of in vitro EC cells into the normal fetal development pathway.

or EK cells, into new strains of mice. The principal challenge may be in devising selection procedures for isolating mutant cell lines in vitro that show interesting developmental Injection into hast blastocvsts (albinagennotype). Ec cells (A gouti genotype) Albino chimaera with Agouti patch phenotypes when integrated into developing embryos. However, the development of direct modification of the germ line through transformation of the early embryo, including the ability to induce mutations with the inserted genes, provides a simpler method of producing genetically modified offspring than the EC cell route.

COAT COLOUR.

The hereditary basis of coat colour is the oldest problem in mammalian genetics. Genetic investigation of coat colour predates the rediscovery of Mendelian genetics in 1900, but the modern analysis of coat colour genetics begins with the work of *Cuenot, Castle* and *Wright* in the first two decades of this century. The work of *Wright* on the genetics of guinea pig coat colour was particularly important; it was the first attempt to use genetic differences to probe the biochemistry of development. The reason for this early attention to the coat colour of rodents is its ready accessibility to genetic analysis. Although the mammalian coat still conceals many of its secrets, the broad outlines of its biology are now visible.

The mammalian coat serves two functions. The primary one is insulation. By constituting an efficient air-trapping layer next to the skin, the packed hairs of the coat provide substantial protection against thermal loss. (The features of birds furnish similar protection.) The coat typically consists of two classes of hair: the shorter and thinner underharis-about 80% of the total hair on a mouse and the longer, thicker overhairs. Within each category, several different hair types exist. The underhairs collectively provide most of the direct insulating capacity of the coat, while the chief function of the overhair'layer *is* to protect the underhair coat. In addition, some overhairs have a specialized sensory role, as do the vibrissae ("whiskers") of the mouse.

The second major function of the coat is to serve as a visual signaling device; in this role, colour becomes crucial. The colour pattern serves as a recognition device for other members of the species, a function that is particularly important in mate selection. In some species, for example, the skunk, it serves additionally as a warning signal to other animals. In some mammalian species, colour serves the function of concealment through camouflage, as the snowshoe rabbit.

All hair shafts develop from multicellular structures termed "follicles", which arise late in fetal development as inpocketings of the epidermis into the dermis. As the epidermal cells push down to form an inverted blunt-ended cylinder, mesenchymal cells of the dermis aggregate at the base of hair bulb. These mesenchymal cells from a thickened, papilla at the *aid* of the follicle that is partially enclosed by the epidennal cells of the hair bulb. It is from the hair bulb that the shaft originates, Each hair consists of an outer cylinder of thin, compressed cells, the cortex, and an internal set of cells, the medullary cells, separated by spaces within the shaft of the hair. Both cortex and medullary cells, but especially cortex cells, are rich in the structural protein keratin. Both cell types are also repositories of pigment. These Melanie pigments are secreted directly into the cells of the hair shaft by melanocytes located in the region of the hair bulb. Pigment is transferred in the form of small pigmentary granules termed "melanosomes" to the basal cells of the shaft throughout the period of hair growth.

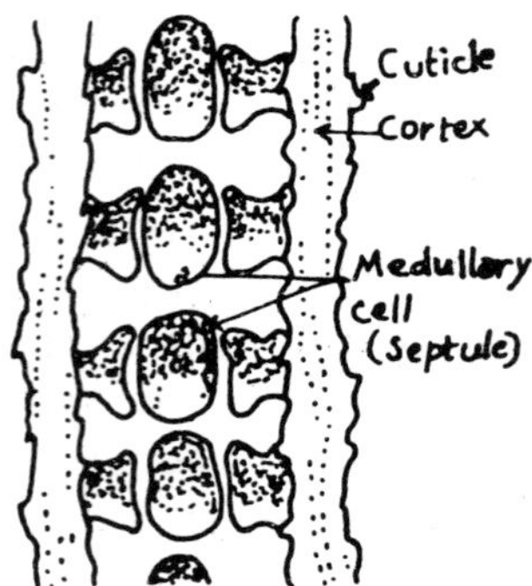

Figure 9.15 : Internal structure of a hair shaft in the mouse coat.

The developmental biology of melanocytes is complex. As discussed earlier, the melanocytes of the skin are derived from the migratory melanoblasts of the neural crest. They arrive at the dermis—epidermis interface at about day 12 of development and form a network of interconnected cells throughout the epidermis. Each active melanocyte eventually transfers its melanosomes either into cells of a developing hair shaft or into neighbouring epidermal cells that are not part of a follicle. The transferred melanosomes are complex entities in their own right each consists of fibrous protein granule attached to a melanic pigment. The pigments are of either of two kinds : eumelanin and pheomelanin. Eumebmin is the prototypical black melanic pigment, while pheomela in yellow. oxide synthesized from tyrosine in a series of sequential ^scarried out by tyrosinase. Pheomelanin differs chiefly from eumelanin in having attached cysteinyl groups. Maturation of the protein moiety of the melanosome takes place gradually, and only mature melanosomes are transferred to epidermal cells. The final colour of a hair of hair subregion is function of which pigment is deposited, the number and spatial arrangement of a melanosomes within the cells of the hair shaft, and the shape and size of the melanosomes.

Given the developmental and biochemical complexity of hair construction, it is not surprising that many genes directly or indirectly exert effects on the colour, colour pattern, or structure of the coat. To date, more than 50 loci have been identified on the basis of their mutant effects on coat colour; a large proportion of these loci have pleiotropic effects and for these, the relationship between colour and other mutant phenes is unknown.

Only a minority of the loci identified on the basis of their coat colour mutant phenotype are directly involved in the synthesis of the melanic pigments. Of the few genes in this category, the albino or *c* locus is the most important. The mutants of the *c* locus form a phenotypic series whose effects range from only a mild diminution of the pigments to a complete absence of melanin. This last is the original c mutant, characterized by a completely white coat and pink eyes. In standard genetic backgrounds, the wild-type *C* allele is dominant to all other members of the series. However, in the presence of certain nonallelic genetic differences that affect coat colour, *C/c* animals can be observed to make less pigment than *C/C* homozygotes.

The albino locus is almost certainly the structural gene for tyrosinase, the key enzyme in melanin synthesis. Not only is there a correspondence between the severity of the phenotypic defect and. the amount of tyrosinase in the allele series, but a particular c, mutant, the Himalayan variety (ce), which is distinguished by a phenotypic temperature sensitivity' in pigmentation—such that the extremities of the animal, which are always cooler, are darker possesses a correspondingly temperature-sensitive tyrosinase activity (Siamese cats, no near relation of the Himalayan mouse, have a similar problem). Unlike the mutants of several other genes whose hair follicles lack melanocytes, the c mutant has normal numbers of melanocytes within the follicles, but they are devoid of melanosomes.

A second locus, *brown (b),* is also involved in the biosynthesis of melanosomes. Mutants of this gene cause the replacement of the black granules with brown melanosomes. The effect of mutations in *b* appears to be on the proteinaceous granules that bind the melanic pigments. When examined microscopically, the melanosomes of *b* mutants are seen to be smaller than those of the wild-type; the visible alteration in colour is a direct consequences of this change in melanosome size.

However, the great majority of coat colour mutants produce their effects on colour indirectly, either through interference with melanocyte development or morphology or through subtle influences on the interactions of melanocytes with the other cell types in the hair follicle. Examples of two genes that affect colour through an effect on melanocyte morphology are the loci dilute *(d)* and leaden *(In)*. Mutants of both genes exhibit an apparent diminution of pigment intensity, the reduction depending upon the allelic composition of the other pigment genes. Thus, homozygosity for *d* causes mice that would otherwise be black to be Maltese blue and those that would be dark brown to be light brown. The dilution effects for mutants of both genes stem from a similar change in melanocyte morphology, although the precise biochemical basis of the change probably differs between *d* and In mutants. Normal melanocytes have a highly branched, or dendritic, morphology that facilitates the transfer of melanosomes. In the mutants, the melanocytes are much more compact. In consequence, melanosome transfer takes place in an irregular, bottlenecked pattern of release. The individual hairs of both mutants show large aggregations of pigment that are unevenly spaced within the hair shafts. Although the macroscopic impression is one of a reduction in pigment, there is as much pigment present as in the wild-type.

A second group of mutants that are indirectly affected in pigmentation are the various white-spotting mutants. Some are dominants, producing the spotting phenotype in heterozygotes; others are recessive. Two of the white-spotting mutants show severe anemias and partial sterility. Although while spotting looks like regional albinism, the cause is fundamentally different; hair follicles in the white areas lack melanocytes. In some mutants, the fundamental defect may be in the neural crest, involving a failure of formation or of migration of melanoblasts. In others, the melanoblasts migrate but die en route. The latter defect has been studied in chimeras and found to occur in the melanoblasts of the mutants, suggesting that the mutation causes autonomous cell death.

Perhaps the most interesting locus affecting coat colour is the agouti gene because of the range and patterns of colour associated with its various alleles. Seventeen alleles of the locus have been identified, ranging from two causing dominantly yellow expression (one, the classic lethal, the other a homozygous viable) to various black or dark brown recessive alleles. Thus, in an otherwise wild-type background, A^y and A^{vy} produce a yellow coat; the A, agouti, standard allele, produces a gray coat: and a^e, extreme nonagouti, produces a completely black coat in homozygots. The agouti locus evidently governs, in some fashion, the balance between phenomelanin and eumelanin synthesis. A general feature of these phenotypes, except for the extreme blackpigmented form, is that hairs on the ventral side are always lighter than those on the dorsal side.

The puzzle of the agouti locus action is revealed by considering the colour pattern of individual hairs in the d A.strain. These hair shafts have black tips, yellow shafts, and black bases. The apparent grayness of the coat in the A strain results from this internal distribution of the two pigments. The transitions between black and yellow in the hair are not abrupt but take place over the length of three to four medullaiy' cells. Indeed, some medullary cells in the transition zone contain both pheomelanin and eumelanin granules. This finding suggests that the gene somehow affects the synthetic behaviour of the melanocyte rather than its intrinsic synthetic capacity for the two melanic pigments.

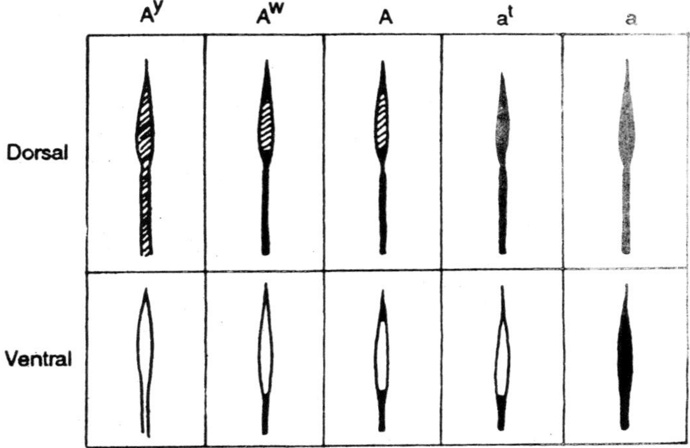

Figure 9.16 : Hair color phenotypes corresponding to representative alleles of the agouti locus.

In fact, the *agouti* locus acts in the epidermal or dermal cells of the follicle rather than in the melanocytes, as shown by the creation of skin chimeras in grafting experiments. When embronic or neonatal skin is grafted between strains, the developing follicles are populated by host melanoblasts. In all cases, the phenotype of the developing hairs, with respect to *agouti,* is characteristic of the genotype of the grafted skin and not of the host melanoblasts.

Thus, if one grafts skin of nonagouti, *ala* (black) to A^y (yellow) hosts, the developing skin graft produces -a patch of black hair, although the hair follicles of the graft are populated with A^y genotype melanoblasts. The results do not reflect the action of an indigenous population of melanoblasts carried along in the skin grafts because the same result ' is obtained when the skin graft is taken from white patches of whitespotting mutants (which lack melanoblasts: the colour phenotype of the developing hair on the graft still follows that of the *agouti* allele of the graft.

Is the instruction affecting pigment synthesis given by the epidermal or dermal cells of the follicle? One can answer this question by making reciprocal dermis-epidermis hybrid grafts using embryonic skin. In such experiments, skin is taken from 13- to 14-day embryos of both *agouti* genotypes and separated into its two components by light trypsinization. The reciprocal hybrid grafts are then made by recombining *the* layers between the two genotypes, allowing a day for the grafts to heal, and then transplanting them individually to the testis of newborn males (a particularly favourable location for growth of the new grafts). If the hair phenotype follows the dermal genotype,one concludes that the gene acts in the dermis. Alternatively, if the phenotype follows the genotype of the epidermis, the latter must be the site of gene action.

The observations have been surprisingly complex. For some agouti allele combinations, the dermis seems to be crucial, but for two others, which produce a "mottling" phenotype (A^{vy} and a^m), the hair phenotype follows the genotype of the epidermis. Poole (1980) has suggested that the immediate instructions to the melanocytes are given by the epidermal cells of the follicle but that the epidermal cells receive their orders from the dermis. In those genotypic combinations in which the epidermal cells govern the hair phenotype, the epidermis may have "set", in other words, lost its competence to respond to a new dermal signal.

The complete set of results shows that the final pigmentation pattern is the product of a subtle interaction between two very different cell types, dermal cells and melanocytes. As wehave seen, these cell types have distinctly different developmental origins. In fact, chimera patterns can reveal such differences in origin and sites of gene action. The melanocytes are ectodermal and derive from the melanoblasts of the neural crest: the major part of the body dermis is mesodermal and derives from the somites. Because the melanocytes and somites have different tissue origins, genes expressed in one of these hair follicle components but not the other show very different spatial patterns of expression from one another in chimeras. All the melanocytes of skin, as we have seen, may ultimately trace their origins to 34 melanoblast clones in the body; what is certain is that the coat patterns of melanoblast chimeras can be reduced to a basic archetypal pattern of 17 stripes on each side. In contrast, agouti-nonagouti (gray-black) chimeras show a different and much more finely striped pattern. Examination of large numbers of these animals suggest that there may be as many as 85 basic stripe widths per side, including about 18 in the head region, altogether a much larger number than seen in the comparable melanoblast-generated patterns.

Two interpretations of the agouti chimera pattern have been offered. The first is that each unit stripe derives from a hair follicle progenitor clone. Apart from the 18 or so in the head, there are about 65 per side for the body and tail, a number that matches the somite number for these regions. In effect, each somite may contain one hair follicle progenitor clone. (In this view, the source of the stripes on the head raises the interesting possibility that there are rudimentary "invisible" head somites, a subject that has been debated by mammalian developmental biologists). The other interpretation of the fine grained pattern is based on the observation that there is much greater regularity of striping in these hair follicle chimeras than in melanocyte chimeras. *McLaren* (1976a) suggested that this regularity signifies a systemic patterning of some kind that is superimposed on the chimeric skin rather than an alternation of genetically determined clones. On the basis of random clonal placement, such regularity is unexpected. That there are *some* systemic influences that regulate *agouti locus* expression is apparent from the consistent dorsoventral differences in pigmentation within single genotypes.

Solution of the various puzzles associated with the *agouti* locus ultimately requires determination of what the gene product does. Earlier results were interpreted to mean that the gene regulates the pheomelanin-eumelanin decision through control of the synthesis of sulfhydryl (SH)-containing compounds, with higher levels of SH compounds favouring eumelanin synthesis. However, this hypothesis has received little subsequent support. The action of the *agouti* gene product remains a puzzle.

Apart from the systemic effects on the expression of *agouti,* the idea that there are general systemic influences on the development of coat colour pattern is inescapable from the fact that many mammals show a species-specific coat pattern of spots or stripes, or sometimes both. Some common examples are zebras, giraffes, and many members of the cat and squirrel families. In contrast to the colour patterns of genetic chimeras, constructed in the laboratory, or those of X inactivation mosaics, each of these species-specific coat patterns is produced in animals whose cells are essentially identical in genotype. The question of the origins of such patterns is therefore one of developmental physiology rather than developmental genetics (although the comparative genetics of the different species' patterns is an interesting matter). In formal terms, these patterning processes can be thought of as morphogen systems acting throughout the presumptive field of pigmentation in the embryo. or fetus. The morphogens may either suppress or enhance pigment formation.

SEX DETERMINATION

Sex Chromosomes and Gonadal Differentiation

The ultimate source of the differences between male and female mammals is the sex chromosomes. As in the fruit fly, females are the homogametic sex, being XX, and males are the heterogametic sex, being XY. Unlike Drosophila, the crucial difference is the possession or absence of the Y rather than the number of Xs. Thus, female XO mice are phenotypically female and fertile (in humans, the XO condition produces sterile females) and XXY animals are phenotypically male, the reverse of the fruit fly situation. Evidently, in mammals, the Y carries one or more sex-determining factors, and maleness is the dominant sexual phenotype (in the biological sense); in the absence of the Y, femaleness results.

There is a further difference in the sex determination mechanisms. In *Drosophila,* sexual phenotype is determined autonomously, cell by cell, whereas in the mouse, gonadal sexual determination is primary, with all other aspects of sexual phenotype following from it. Secretion of the hormone testosterone by the male gonad triggers male genital development; in the absence of testosterone, the gonad secretes estrogens and the secondary sex characteristics that develop are female. This mechanism is general in mammals.

The first differences in gonadal differentiation become apparent at about 12.5 days p.c., 2 days after the arrival of the first PGCs. The male gonad begins to organize itself into testis cords, consisting of solid strings of spermatogonial cells encased in mesodermal somatic cells. The early female gonad retains a compartmented appearance, with groups of oogonial cells surrounded by a matrix of mesodermal cells. The urogenital system undergoes an accompanying differentiation. Initially, both sexes possess both kinds of urogenital structures, the female Mullerian ducts and the male Wolffian ducts. In each sex, one duct system develops while the other degenerates.

Although the primacy of gonadal differentiation in setting sexual phenotype is dear, it is not known with certainty whether the initial events occur in the gonadal soma or the germ line. In the chick, it has been possible to show that the soma of the gonad plays the crucial role: when PGCs are prevented from colonizing the genital ridge, the gonad nevertheless develops according to the sex chromosomal composition of the (somatic) ridge. In mammals, this kind of unequivocal demonstration has not yet been possible. What can be done is to construct chimeras, and analyze for correlations between phenotypic sex and the respective chromosomal composition of the somatic and germ line tissues of the gonad. Because this procedure almost never produces an animal that has reverse compositions for the two components, the interpretations are always uncertain.

However, the chimera results show two things. The first is that most the XY/XX chimeras (which comprise about half of the experimental animals) are male, a consequence-of the dominance of maleness. Above a certain threshold of male (XY) cells, perhaps as little as 30%, a chimeric gonad tends to develop as a testis, while around that threshold, an ovotestis (a mixture of testicular and ovarian material forms. Development of a testis then brings in train the development of secondary male sex characteristics.

The other finding of interest is that in XY/XX chimeras, regardless of the phenotypic sex of the gonad, there is little or no conversion of germ line cells of the "wrong" composition into functional gametes. For XX germ line cells developing within a testis, the prohibition seems to be absolute and may reflect an absolute block of spermatogonial development in cells with two X chromosomes. For XY cells, transfer of germ line cells out of the somatic gonad before these cells encounter the normal meiotic block typical of the developing male gonad can sometimes permit oocyte development. Possession of the Y chromosome, therefore, is not an intrinsic block to oocyte development.

The probable sequence of events therefore is an initial commitment in the somatic portion of the gonad created by the sex chromosome composition of that gonad, followed by a distinctive course of gamete maturation, and finally by the entrained development of secondary sex characteristics by the secreted male or female. sex hormones.

Mutants and the Role of the Y Chromosome

Sexual development in animal systems with sex chromosomes involves two stages, an initial phase of sex determinative decisions and an ensuing period of sexual differentiation. Both sets of processes are subject to genetic derangement. Because of the temporal separation between the initial event (gonadal determination) and the subsequent differentiation events, it is possible to male informed guesses as to which process has been genetically altered.

In mammals, the differentiation period is dependent on the continued presence of the secreted sex hormones: if androgen production is stopped by removal of the testes in male fetuses, the further course of somatic sexual development becomes shifted toward the production of female characteristics. Similarly, mutations that interfere with either gonadal sex hormone production or hormone binding would affect the later stages of sexual development.

Although mutants deficient in hormone production have not been reported, a sex differentiation mutant, characterized by insensitivity to testosterone, has been described. It is the X chromosomal testicular feminization *(Tfm)* mutant, and its phenotype is that of partially feminized males. (The mutation has no effect in females). XY mice that carry *Tfin* develop testes, but these are small and underdeveloped; externally, the animals resemble females, although not perfectly. The animals have normal testosterone levels, and administration of neither testosterone nor dihydrotestosterone rescues the male phenotype. Evidently the defect is not in testosterone production but in the response to testosterone. Similar genetic syndromes have also been reported in rats and humans. In all cases, the specific defect appears to be a deficiency of the androgen receptor protein, which is normally distributed in all or most cell types.

The most dramatic effects of *Tfm* are on secondary sexual characteristics, but there is also a block to spermatogenesis. In *T fm/Y* individuals, spermatogonial are prevented from giving rise to spermatocytes, the immediate precursor cells of the sperm. To determine whether the effect is a direct one in the germ line or a secondary effect of an inappropriate gonadal soma, Lyon et al. (1975) constructed chimeras of *Tfm/Y* and +/ *Y* embryos and then mated the mature chimeric males with *Tfm*/ + females. Some of the progeny obtained were found to be *Tfm/Tfm* embryos, showing transmission of some Tfmcarrying sperm from the male chimeras.

The result indicates that the presence of some *Tfm*$^+$ in the gonadal soma can rescue spermatogenesis in some of the *Tfm* spermatogonia, and that the effect of the mutation on spermatogenesis is in the soma rather than the germ line.

Sex determination mutants should differ from sex differentiation mutants in having transformed gonads as well as transformed secondary sex characteristics. Although hunts for sex determination mutants in mammals are infeasible, because large numbers of animals cannot be screened, a number of putative autosomal and X chromosomal sex determination mutants have been identified in mice, goats, wood lemmings, and humans. In mice, several genetic conditions that affect sex determination have been identified. However, the precise genetic basis of most of these conditions remains ill-defined, including the number of loci involved. For several, the effect involves an aberrant interaction between the Y and the autosomes, where the Y has been introduced from other subspecies or lines, resulting in a partial f eminization of XY individuals. The genetic basis of these syndromes is obscure, but their existence shows that simple possession of a Y is insufficient to guarantee normal male development-that Y-aautosome interaction in some manner is essential for this development.

Of the several murine genetic conditions. that alter sex determination, that of Sex reversal (Sxr) is the best understood. It was first described by Cattanach et al. (1971) as a dominant mutation that transforms Xx animals into sterile males. Because of this sterility the initial mapping had to be carried out in crosses between normal females and XY Sxr carrier males; the results indicated that the condition was inherited independently of both the X and Y chromosomes. In consequence, the authors concluded that the trait was autosomal. However, all subsequent attempts to localize it to a particular autosome met with failure.

The solution of the *Sxr* puzzle was made possible by the use of a molecular marker specific to the Y chromosome. This marker is a particular satellite DNA that is unique to the Y in mammals but was first isolated not from mice or humans but from a snake, the banded krait. More surprisingly, it was isolated from the heterochromatic female-specific W sex chromosome of snakes. In the banded krait, as in most snakes,

sex determination is carried out by the ZW-ZZ sex chromosome system. In this system, male are the homogametic sex, with two euchromatic Z chromosomes, and females are the heterogametic sex, with a ZW constitution. By looking for a satellite DNA species unique to the heterochromatic W chromosome, *Singh* and *Jones* (1992) isolated such a satellite, which they designed Bkm (for "banded krait, minor satellite").

The Bkm satellite is found throughout eukaryotes, from yeast to humans although in varying amounts and different locations. The fact that it is found on the female-specific chromosome of snakes and on the male-specific chromosome of mammals implies that it does not determine femaleness or maleness per se but that it is a marker of sex chromosomes. When radioactively labeled Bkm satellite is hybridized in situ to metaphase spreads of chromosomes from XX *Sxr* animals, it is found to be located at the distal tip of one of the X chromosomes. The result indicates that the *Sxr* condition derives from the translocation of Y chromosomal material to the distal tip of the X, presumably by recombination during meiosis. The transfer of Y chromosomal material to the X by a highfrequency recombination event explains the absence *of* genetic linkage between the Y and the Sxr condition found in the earlier studies.

Eicher (1993) has proposed a hypothesis for the event. The idea is that the *Sxr* condition is characterized by a Y chromosome carrying two sex-determining, Bkm-containing regions, and

that transfer of the more distal region from one chromatid to the X invariably occurs during meiotic pairing in the XY *Sxr* animals. Note that the process, as diagrammed, produces equal numbers of Xsx' and normal Y chromosomes. Indeed, XY Sxr carriers give rise to normal males in addition to transformed XX females and carrier XY *Sxr* sons (who bear the nonrecombinant, duplication-bearing chromatid).

The central question about mammalian sex determination concerns the manner in which the Y chromosome confers maleness.

The general belief is that the Y carries one or more genes that actively promote testicular development. Much speculation has focused on the possibility that one of the known male-specific antigens might play that role and be encoded by genes on the Y. One of these is detectable as a transplantation antigen, causing the rejection of male skin in females of the same inbred strain. This antigen has been dubbed the "H-Y antigen (for the Y histocompatibility antigen) and is both highly conserved and male specific throughout the vertebrates. The possibility that the H-Y antigen is encoded by a gene on the Y and is the male-determining factor was originally proposed by Ohno (1986). A second antigen that is male specific was subsequently discovered in sera and designated SDM. For a long time, it was believed that SDM and H-Y were the same antigen. It is now apparent that they are not. More importantly, the' current evidence shows that neither H-Y nor SDM is invariably associated with phenotypic maleness, nor is their absence invariably correlated with femaleness. These antigens are therefore probably consequences rather than causes of male determination. The nature of the relationship of the putative male-determining factors to the Bkm satellite also remains to be elucidated. The Bkm sequence may contain protein coding functions but these are presumably not directly male-determining because Bkm is not invariably associated with maleness in vertebrates. If there are male-determining, protein-coding function on the Y, they are presumably closely linked to the Bkm satellite.

A different view of the role of the Y in setting maleness has been proposed by Chandra (1994), who suggests that it is purely passive: to absorb repressor molecules, thereby permitting the function of an X chromosomal gene essential for testicular determination, designated Tdm. In this view, the Bkm satellite might be the male determinant itself by functioning as the repressor binding sequence. Chandra's specific proposition is that an autosomal locus encodes a repressor that binds with high affinity to the repeated Y chromosome sequence and with lower affinity to Tdm. In animals possessing one or more Y chromosomes, the repressor is bound by these chromosomes and Tdm expression is turned on; in XX animals, there is no sink for repressor and Tdm is repressed. Although this model is at best an approximation, it fits most of the known facts and provides a provocative alternative to the conventional view of the Y chromosome's role in sex determination.

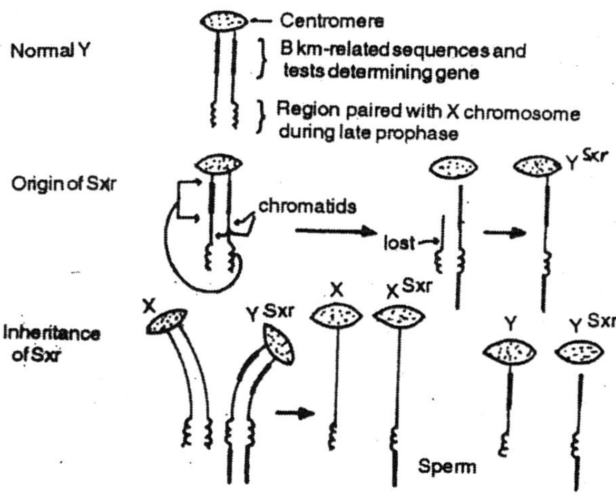

Figure 9.17 : Model of the origin of Sxr-carrying X chromosomes.

Superficially, sex determination in mammals appears to operate by *very* different rules than in fruit flies or nematodes. The controvertible difference is that in fruit flies, the decision is made autonomously in all cells capable of a sexually dimorphic response, while in mammals the key sex determination events occur in one organ, the gonad, with all of the remaining sexual dimorphism following from this first step. There may, however, be certain underlying unities. In both organisms, the decision is a binary one in the cells that make it-either a "yes" or "no" to development along the pathway of the dominant sex. In addition, the key sex chromosome (the X in fruit flies, the Y in mammals) interacts with one or more autosomal factors in making the sex determination decision. These autosomal genes have been identified in *Drosophila,* while those of mammals are still obscure. However, the general logic of such ait interaction, if not the precise molecular details, may prove to be surprinsingly similar. The future of this problem is certain to be interesting.

10

Human Development

SCARCITY OF EARLY HUMAN EMBRYOS

Because of the inaccessbility of the fertilized human ovum, its cleavages have never been observed. Human embryos can be obtained only through accidental or surgical abortion from the uterus, and such embryos are past the cleavage period. Even later stages are comparatively rare, since *they* are extremely small and are usually lose or injured in scraping the uterus and searching for them. However, of late two authors, *Hertig* and *Rock* have published a great deal about early human embryos. One of these concerns an embryo nine days old of which a reconstruction has been made, and one is even younger. These two men have added greatly to our knowledge of human embryology.

DEVELOPMENT OF BLASTOCYST

Since the early cleavages of mammalian ova have been found to be uniform in their cleavage patteut, it seems reasonable to assume that the human egg is no exception in this respect. At any rate, one may safely say that its cleavages are like those of the macaque monkey. The arrival of the blastocyst in the fundus of the uterus has been variously estimated to be from four to seven days after fertilization, but it has been established that the tubal journey of the macaque monkey takes *only* three days and that would confirm the lesser estimate of four days in man. From the study of the estrous cycle it is safe to assume that it arrives within

4 to 6 days after ovulation. Basing our conclusion on the development of the blastocyst in other

mammals, the human embryo at that time is probably slightly past the morula stage. The inner cell mass, or embryonic knob, has been formed and the trophectoderm consists of a single cellular layer functioning as an envelope of the blastodermic vesicle. After adhering to the uterine wall for a few hours, it begins to sink into the utterine mucosa or endometrium by localized cytolysis. Apparently the erosion of the tissue is due to the cytolytic action of the trophectoderm cells. This destructive action of the blastodermic vesicle has been observed in several mammals, as for example the guinea pig. Further more, it appears that the blastocyst elongates in its descent into the endometrium, thus temporarily losing its spheroidal shape. Having disappeared beneath the surface of the uterine mucosa, it readjusts itself, resuming more or less its former contour.

The presence of the blastodermic vesicle within the endometrium has a destructive effect on the uterine mucosa immediately surrounding it. Indeed, this deleterious action of the trophectoderm is even more pronounced than the one taking place. during its entrance into the endometrium. The cells of the trophectoderm multiply at a rapid rate and push into the adjoining *uterine* tissue. The latter is broken down under the cytolytic action of the trophectoderm cells and forms cellular fragments, broken-down capillaries, fibrin, degenerating blood cells, and other cellular debris. This material is called *embryo-troph and* is absorbed by the blastodermic vesicle to supply the necessary food f growth. This embryotrophic nutrition lasts until the embryo has developed its own vascular system and perfected its placental attachment. One of these early human embryos was described by *Hertig* and *Rock* in 1945 and is represented in which is based on a plastic reconstruction of the uterus and its glands.

The original trophectoderm has now considerably increased in size. Its inner-most layer remains distinctly cellular and is from now on referred to as the *cytotrophoderm (cytotrophoblast)*. However, the cells which have been proliferated from it and have invaded the uterine tissue from syncytia and loosely organised spongy tissue called the *plasmotrophoderm (plasmotrophoblast)*. The latter acts as does a parasite, disintegrating and living on the tissue of the host. Spreading and increasing its destructive effect, the plasmotrophoderm comes in contact with uterine blood vessels : it disintegrates their walls and maternal blood comes in direct contact with the plasmotrophoderm.

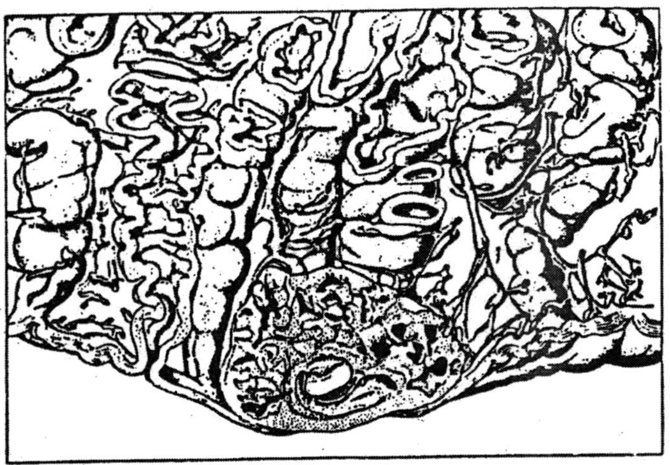

Figure 10.1 : Reconstruction of the 9-day ovum, sectioned through the middle, viewed at right angles to the cut surface.

At that stage of its development, the cytotrophoderm attains a second period of rapid cell proliferation which is followed by a corresponding diminution of the plasmotrophoderm. This new cellular multiplication produces minute projections or *chorionic villi*. As the plasmotrophoderm shrinks, the chorionic villi come in direct contact with the blood and form in this manner the foundation of the discoidal, deciduate placenta.

EARLY HUMAN EMBRYOS

Hertig and Rock (1945) described a human embryo, the age of which they estimate to be seven days. If one assume that the segmenting egg arrives in the fundus of the uterus after four days of moving down the oviduct, then this 7-day embryo is not much older than the blastocyst.

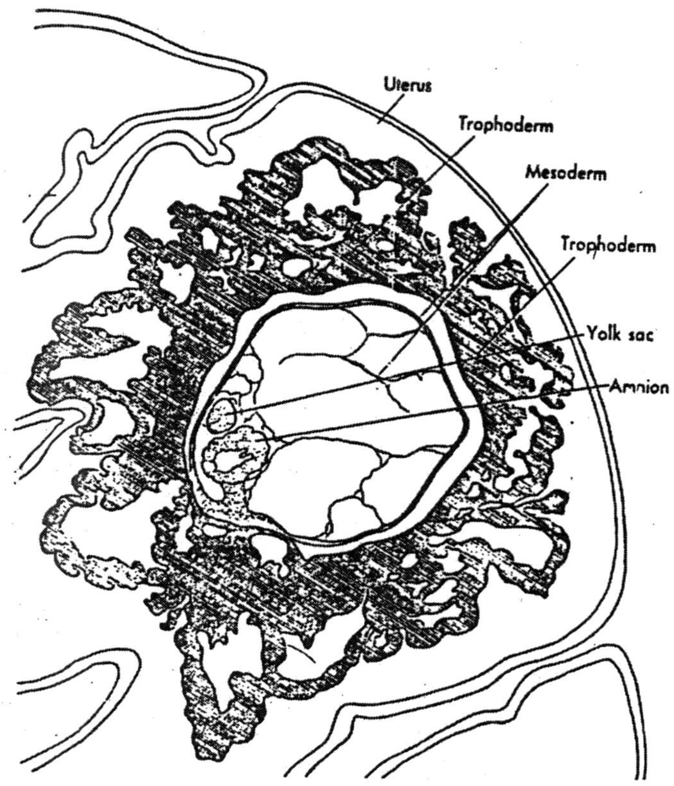

Figure 10.2 : A human embryo.

A blastodermic vesicle of a slightly more advanced stage has been described by Streeter in the Miller ovum. It has been estimated to be from ten to twelve days old. The cytotrophoderm and plasmotrophoderm are indicated by the heavily-shaded area. The mesodermal layer is shrunk away from it, and fine mesodermal strands are seen all over the vesicle. The amnion and the yolk sac have been differentiated, but the allantois is not apparent as yet.

At this stage of development the future embryo can be allocated to the area represented by the embryonic shield. It appears that the blastocyst has grown extensively by the increase of the cytotrophoblast and its chorionic villi. The condition of the young embryo indicated in the diagram is realized in the Peters embryo as well as in von Mollendorff's embryo. The Peters embryo was probably the best preserved as well as the most interesting of all the human embryos up to the description of such embryos by *Hertig* and *Rock*, and even these two writers give it adequate credit in their pages. The embryo is estimated to be fifteen days old, measuring 0.19 mm. in length. It appears that in this embryo

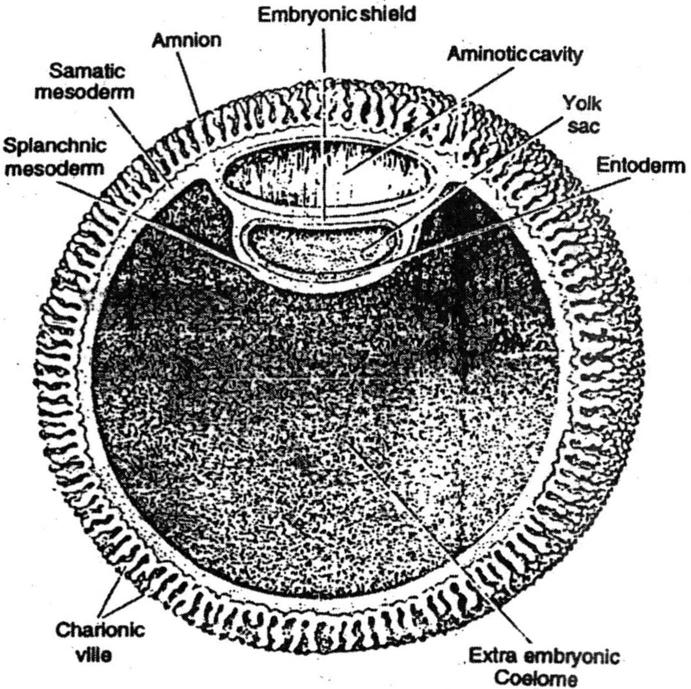

Figure 10.3 : Stereogram of a human blastcyst contianing an embryo of approximately fourteen to fifteen days.

the plasmotrophoblast has become reduced and that the cytotrophoblast has begun its second period of cellular multiplication. Chorionic viii extend peripherally into blood lacunae which are in direct contact with blood vessels. These lacunae become the large blood sinuses of the placenta. The chorionic villi with their central cores of mesoderm from nipple-like projections from the cytotrophoderm. Here and there one may be seen to form secondary beaches which will finally culminate in the complex chorionic villi

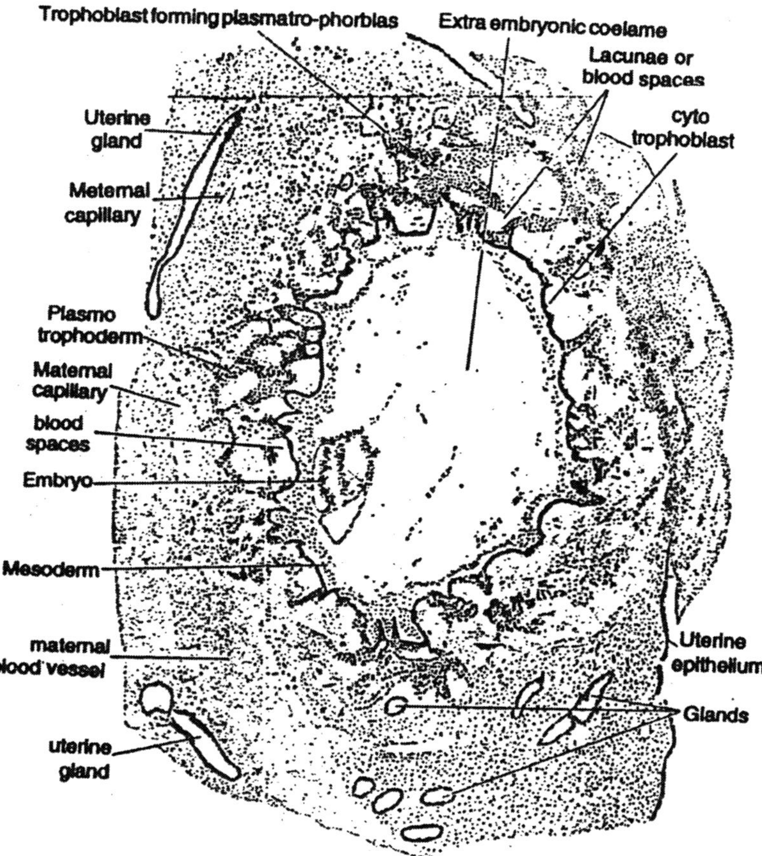

Figure 10.4 : Human embryo estimated to be fifteen days old and 0.1 mm. long.

of the mature placenta. Though the extent of the plasmotrophoderm has been reduced, the destructive action against the uterine mucosa is still continued and is now carried on by the cytotrophoderm by means of the rapidly extending and branching chorionic villi. Nutrition is still embryotrophic, but the beginning of the hemotrophic form of nutrition is in preparation.

The triangular space on the left side has been tentatively diagnosed by Peters as the extraembryonic coelome. He is not certain whether this identification is correct. It may merely represent a vacuolized area. surrounded by a more distinct. mesodermal stand, of which many are visible in the cavity of the large extraembryonic coelome.

Von Mollendorff's embryo OF is probably a day older than that of Peters. Its chorionic villi are larger and more branching, and their rapid growth has pushed them much deeper into the blood lacunae.

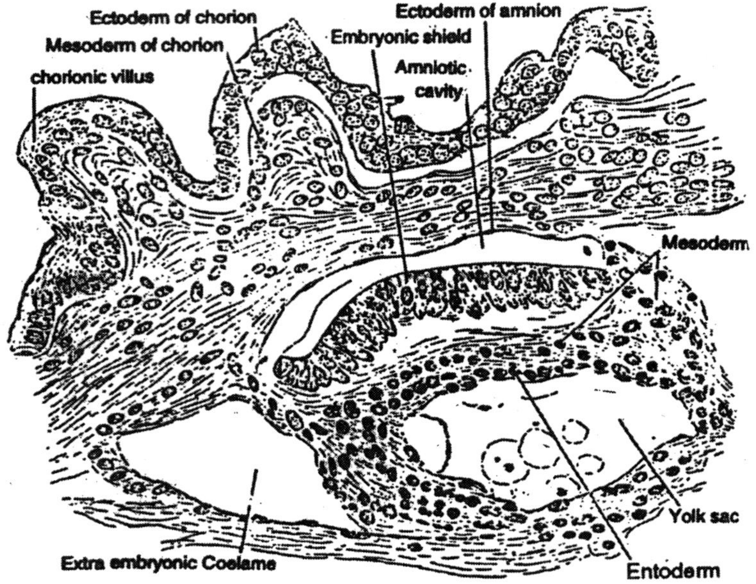

Figure 10.5 : Human embryo.

In the next stage of development a radical change occurs in the position of the embryonic mass in that it detaches itself partially from the inner surface of the cytotrophoderm, of chorion, as it *may now* be termed. At the same time, it increases rapidly in size and forms a tabular outgrowth from the upper region of the yolk *sc into* the body stalk. *This tubular* extension is the diminutive allantois, which grows toward the chorion but does *not spread* over it as it does *in* the ungulates or carnivores. The Mateer embryo described *by* Streeter represents the initial stages of this change. This Mateer embryo is estimated to be seventeen days old and has a dimension of 0.92 mm. It is therefore five times larger and only two days older than the Peters embryo. It is just detaching itself from the chorion, some of its connecting fibres being still intact between the amnion and the chorion. Viewed from the dorsal aspect with the amniotic membrane removed, the Mateer embryo shows a primitive streak on the embryonic shield contianing an anteriorly located primitive pit.

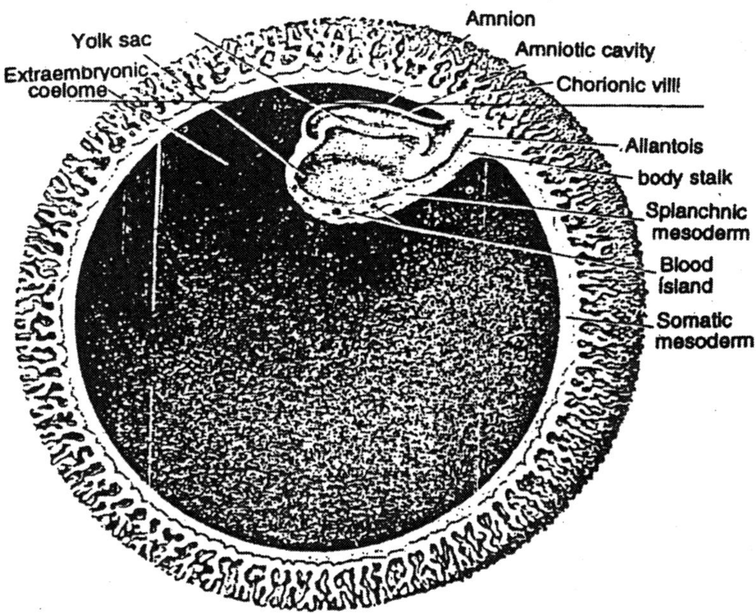

Figure 10.6 : Stereogram of a human blastocyst containing an embryo of approximately twenty to twenty-two days.

The embryo of *Graf von Spee,* which he terms embryo Gle., is estimated to be twenty-one days old and is 1.54 mm. long. It has a well-developed yolk sac containing blood islands for the organization of the vitelline circulation. The fact that it is nearing the end of its embryotrophic nutrition is indicated by the profusely branching chorionic villi and their close association with the blood sinuses of the endometrium, in which they are suspended and surrounded by maternal blood. The chorionic villi are becoming vascular and will soon join the umbilican circulation which is established within a few days. The primitive pit observed in the Mateer embryo and a few other human embryos has disappeared in the Graf von Spee embryo, as was established in later investigation.

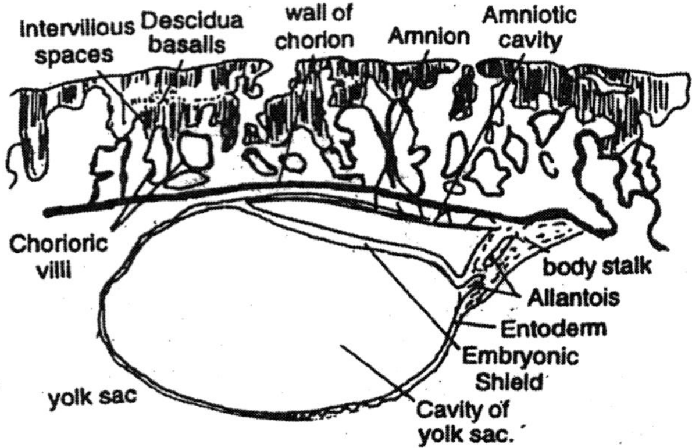

Figure 10.7 Human embryo estimated to be seventeen days old and 0.92 mm. long.

An instructive and illuminating contribution by *Hertig* and *Rock*, featuring human embryos of different ages, is contained in the accompanying illustration. The legend is self-explanatory and includes several of the embryos discussed in this chapter. They are estimated to be from 11 to 15-16 days old.

Between the sixth and seventh week of gestation the embryo grows so rapidly that the blastodermic vesicle forms a bulge on the surface of the endometrium. It pushes the uterine lining (decidua) ahead of itself in its extension into the uterine cavity. During this period the amnion also increases in dimension so that the extra-embryonic coelome becomes almost entirely replaced by it. This expansion of the amnion is the cause for the change in the position of the embryo. It rotates almost 180° from its original position, and now faces the body stalk with its venal side. When the expansion of the amnion is completed its envelops the body stalk and becomes opposed to the inner side of the chorion. The body stalk is from then on known as the umbilical cord.

The membranes have been cut away to expose the embryo. The chorionic villi have progressively diminishes in size and number on the side facing the uterine cavity. In later stages the chorionic villi are relegated to the uterine side of the chorion to form the

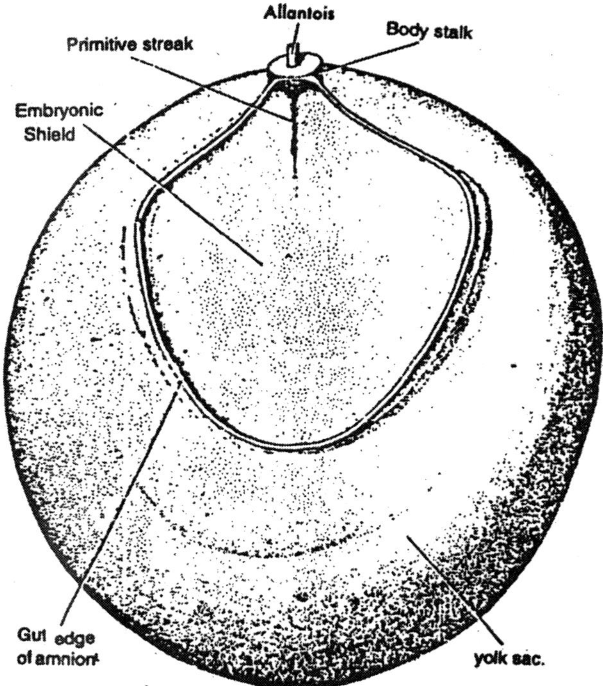

Figure 10.8 : The same embryo as shown in dorsal aspect. The amnion has been cut away to expose the embryonic shield.

disc-shaped or discoidal placenta. As the embryo grows away from the placenta, the two remain in communication by the umbilical cord. The yolk sac and the allantois have remained vestigial organs throughout the entire development; they have become secondary in their importance in the development of man.

PRENATAL INFLUENCES

It was formerly assumed that the mammalian embryo develops entirely independently of its mother, that its embryology was the function of the embryo alone, and that the parent merely supplied the "raw materials" in the form of food, protection and warmth for its development. This belief is no longer held, for with the development of modern chemistry and the better understanding of semipermeable membrance, much progress has been made in

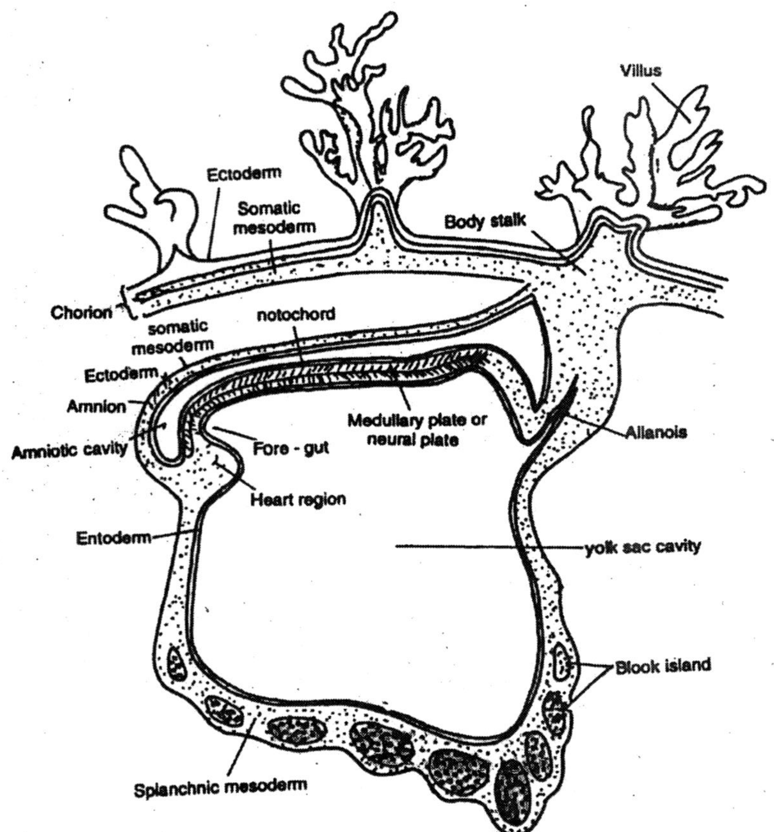

Figure 10.9 : Human embryo described by Graf von Spee, estimated by him to be twenty-two days old and 1.54 mm. long, and referred to as "Ovum Gle".

prenatal care of the human embryo. These investigations show that a good many substances can be passed from mother to fetus or from fetus to mother in the ordinary metabolic processes of the embryo. Thus, it has been demonstrated that certain substances, such as salts, vitamins, hormones and other, can be diffused through the placenta into the fetal blood. This is to be expected since, in the natural course of metabolism, gases and food substances and excretory products in solution have to pass through the placenta. These facts have been established by feeding the mammalian mother certain substances and examining the blood of the fetus

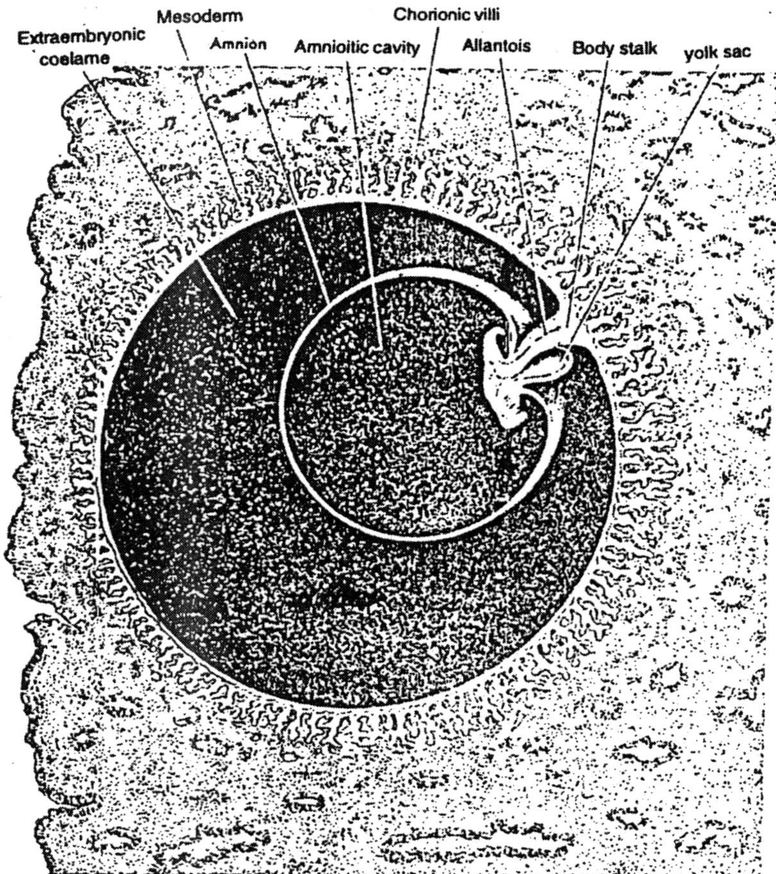

Figure 10.10 : Stereogram of a blastocyst containing a human embryo, twenty nine days old and 4.3 mm. long. The blastocyst is bulging slightly over the uterine mucosa into the uterine cavity. The embryo has rotated and is facing the body stalk. The extraembryonic coelome has been partially replaced by the amniotic cavity.

afterwards. Some materials in solution do not pass because of the selective action of the placental membrane, but a great number do enter the fetal circulation.

However, there are also many instances of large molecules, as for example, proteins and even blood cells passing through the placenta when it has been injured by inflamation or weakened mechanically. Thus a uterus that is infected can be very troublesome for the embryo and may cause its abortion or death. Certain infectious and communicable diseases, as for example syphilis, may easily pass from mother to offspring and do considerable damage there through an infected placenta. Within the last few years, a considerable number of such cases have been recorded, one type of which, the cause for certain still-births, is best established. This is the well-known fetal *erythroblastosis,* a rather common cause of death of the infant before birth which is due to the destruction of its blood and the blood forming centres. To understand the cause of this disease, with all its ramifications, involves the knowledge of immunity reactions and also the laws of Mendelian inheritance.

The greater majority of human carry a factor in their blood which agglutinates the corpuscles of certain other human blood. This is due to the presence of the Rh factor, Rh standing for Rhesus monkey, in which this factor was first discovered. Subsequently, the same factor was demonstrated by *Iandsteiner* and *Wiener* in human blood. With regard to the latter, there are two types : the Rh+, having the Rh factor, and the RH-, having none of it. Being a dominant Mendelian factor, the Rh factor shows individuals who are heterozygous with regard to the Rh factor. An Rh—mother may conceive form an Rh+ father a child which is Rh+ (Rh+ being dominant). When the placenta is defective for one reason or another as shown above, or is so for some other cause not known to us as yet, a limited but continuous amount of fetal blood will enter the mother's circulation and cause the development of anti-Rh substances in the mother, which then pass back to the embryo by means of the fetal circulation. In the fetus these anti-Rh substances will harm its blood. If there are not too many such substances generated and passed over by the mother, the child may live; but in the second and further pregnancies, there is an increasingly high danger that the child will be still-born, because with every pregnancy more of the anti-Rh substances are developed so that eventually there will be enough to kill the infant.

MULTIPLE BIRTHS

Within recent years the interest of the public has been centred on twinning and multiple births. They are of common interest because they are considered to be comparatively rare in man, and because' the similarity of certain twins is so striking that it arrests the attention of the crowd.

It is a familiar fact that many mammals give birth to more than one young at a time, and that others, usually the larger ones, being forth only one at birth. One may make the arbitrary rule that this phenomenon is due to the size of the animals. For example, such mammals as the horse, elephant, and the giraffe give birth to only one young, while dogs, cats, rabbits, mice, and others produce litters

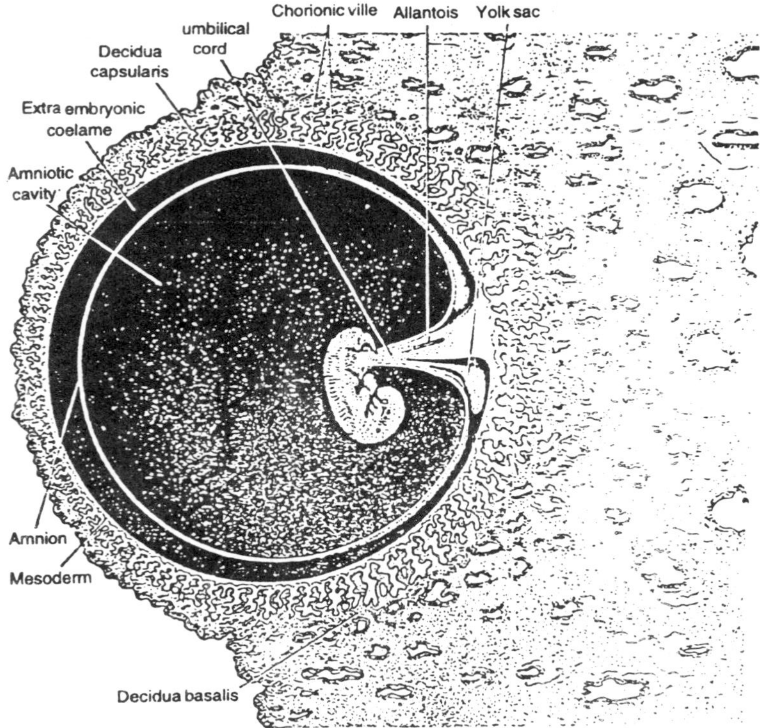

Figure 10.11 : Stereogram of a blastocyst containing a human embryo, thirty-three days old and 5.00 mm. long. The blastocyst has formed an elevation on the surface of the uterine mucosa. The extraembryonic coelome has been almost entirely replaced by the amniotic cavity. The body stalk has been converted into the unbilical cord by being covered with the amnion.

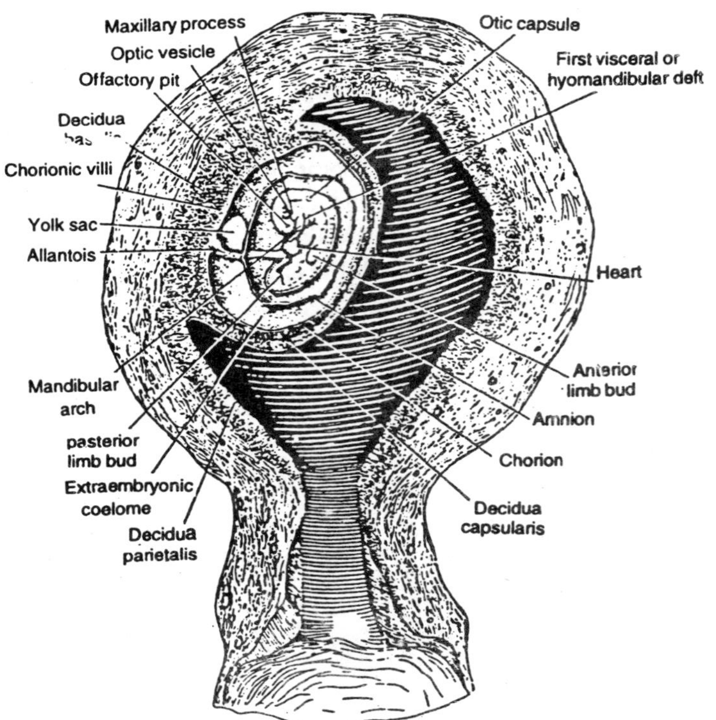

Figure 10.12 : Human embryo in the eterus, thirty-eight days old and 7.5 mm. long. (Adapted from Coste and other sources.) The chorion and amnion have been cut away to expose the embryo to view. The yolk sac extends into the extraembryonic coelome. The vestigial allantois may be seen next to the yolk stalk. The umbilical arteries and veins are not shown.

of many individuals. That this division is not universally true in exemplified by the lion, which is larger and heavier than man and may have a litter of several cubs, or by the pig, which may bright forth a dozen young at one time. Large litters are chiefly confined to mammals having a bicornuate or bipartite uterus, but even this condition has its exceptions.

Twins, triplets or other multiples may be formed in three different ways in man : they may arise from two independent ova, or they may develop from a single one, or they may be formed by a combination of the two. In the first instance they are referred to as *fraternal* twins because they are not any more closely related

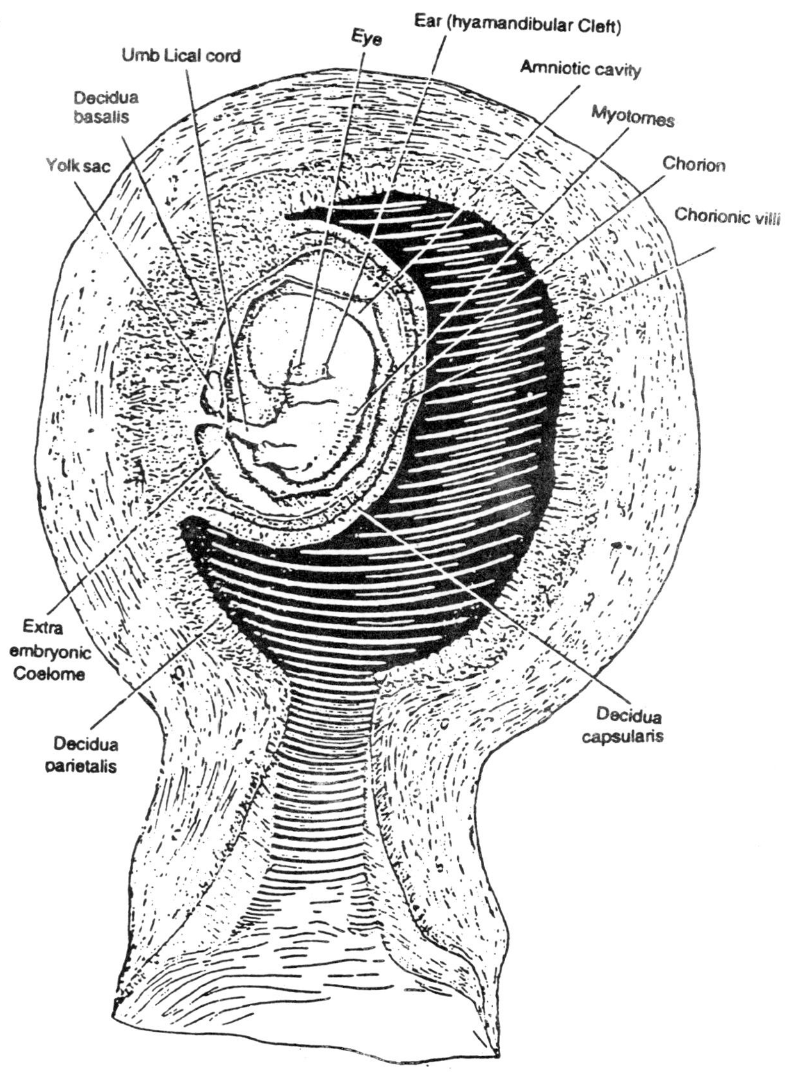

Figure 10.13 : Human embryo in the uterus, seven weeks old and 17 mm. long.

to each other than they are to other brothers sisters in the family. Having their origin from two different ova, the resulting blastodermic vesicles are entirely independent and develop their separate fetal membranes. In later development the placentae may approach each other and may even grow together, but their separate identify can always be demonstrated by the independence of the umbilical blood circulation. Fraternal, or more accurately *biovular,* twins have a much chance to be of the same sex as to be of different

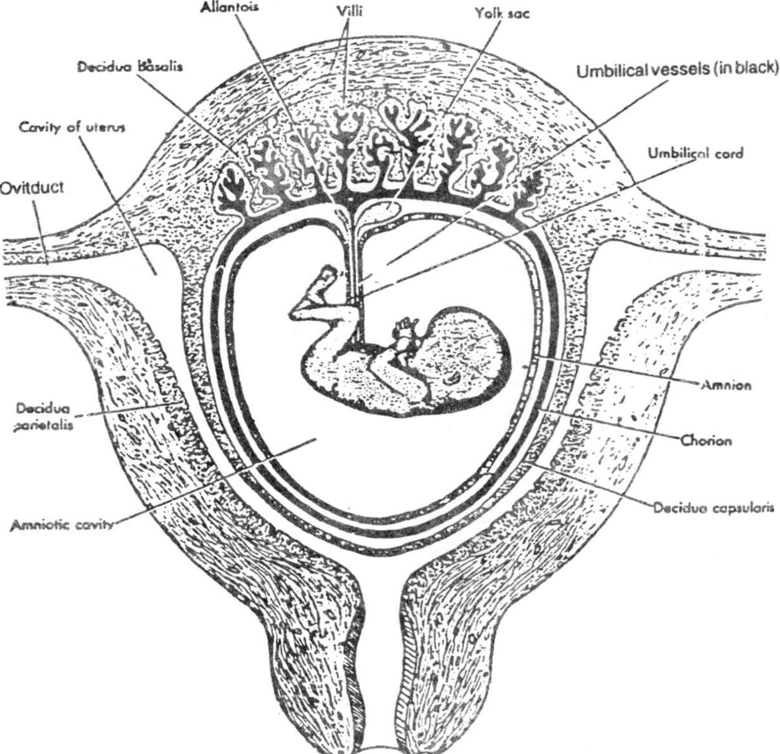

Figure 10.14 : Sectional diagram of human uterus with fetal membranes and their relationship to the uterus and the embryo.

sexes. Their origin may be due to the simultaneous ripening of two Grafian follicle and the liberation of two ova for fertilization. Another cause of their origin can be ascribed to the presence of two oocytes within one Graafian follicle. In either case biovular twins will be formed, and the possibility of producing triplets and more individuals is determined by the increase of these causes or their combination.

Those of the second type, called *identical or monovular* twins, are always of the same sex, and they resemble each other so closely that it is sometimes difficult to tell them apart. Monovular twins are derived from the same zygote. Their duality is established early in the formation of the blastodermic vesicle. From evidence gathered in other mammals, such as the pig and e specially the armadillo, it seems that the inner cell mass or possibly the embryonic shield divides into two parts, each developing a separate amnion, but retaining the same yolk sac, chorion and placenta. In the formation of triplets, quadruplets or quintuplets of this second type, the inner cell

mass separates into the corresponding number of units. Since monovular twins have identical chromosomes and genes, they are identical in their physical and mental traits and characteristics. There is no closer human relationship than between identical twins. They are more closely related to each other than to their father and mother or any other member of the family.

A third and more common method by which triplets and other multiples are produced is by the combination of the two processes discussed above. Such multiples show this dual formation usually in the sex ratio, placenta, and in their general appearance.

Twinning is not quite as rare as it is generally assumed. One birth out of every eighty-five will have twins. Triplets occur in one out of 85 2 (=7.225), quardruplets in one but out of 85 4 = 32,2000.625) > Mortality of infants of multiple births increases in direct ration of multiplicity. The higher the multiple, the less chance there is for the individual to survive. The survival of quintuplets is a biological phenomenon.

11

Hormones in Development

The normal human menstrual cycle is dependent upon a complex system containing several components, including higher brain centers, the hypothalamus, the pituitary gland (hypophysis), the ovaries, and the uterus. The cohesive function of these components is integrated by positive and negative feedback signals, and comprises the endocrine system in the female.

In the male the same system operates, only the testes take the place of the ovaries as the major steroid-hormone-secreting organs. Briefly, the reproductive endocrine system functions as an integrated unit in which a releasing hormone (known as gonadotropin releasing hormone [GnRH], or luteinizing hormone releasing hormone or factor [IRF or LHRH] is produced and released by the hypothalamus under the stimulatory influence (positive feedback) of steroid hormones produced by the target organs (ovaries and testes). LRF then stimulates the anterior pituitary gland to release the gonadotropic hormones-follicle-stimulating hormone (FSH) and/or luteinizing hormone (M-which stimulate production of steroidal hormones from the target organs. In the case where steroid hormones inhibit release of hypothalamic releasing hormone, this is known as negative feedback (see the section on Integration).

Recent technological advances which have facilitated the

measurement of hormones in biologic fluids, coupled with advances in the study of the interaction between hormones and target organs, have opened the way to a better understanding of the regulatory systems. We now enjoy a reasonable complete understanding of the endocrinology of the menstrual cycle. This chapter will first discuss the individual components of the system, then their integrated function, and finally the effects of the hormones on their target organs.

THE COMPONENTS

The Central Nervous System

Although there are conflicting data and hypotheses, the majority of evidence suggests that transmitter substances such as dopamine and norepinephrine, in the brain (particularly in the hypothalamus) affect the specialized nerve cells which contain the hormone-releasing and inhibiting factors. Current data support the existence of a dual system: dopamine inhibits, and norepinephrine stimulates, gonadotropin releasing factor secretion. The cell bodies of the neurons containing norepinephrine are largely located in the medulla and pons, whereas those containing dopamine are found in highest concentration in the arcuate and periventricular nucleii. This type of neuronal activity is largely regulated directly by changes in the circulating levels of the steroid hormones—estrogen and progesterone.

The hypothalamus secretes a releasing factor for each of the tropic hormones produced by the anterior pituitary gland except the gonadotropins, which are probably regulated by a single releasing hormone. The structures of two of these hypothalamic releasing hormones, thyrotropin releasing factor or hormone (TRF or TRH) and gonadotropin releasing factor or hormone (LRF, GnRh, or LHRH), have been elucidated. Thyrotropin releasing factor is a tripeptide that causes release of thyroid-stimulating hormone (TSH), as well as prolactin (PRL)—another trophic hormone which has a major action on the mammary gland. However, it may not be the primary prolactin releasing factor. Gonadotropin releasing factor is a decapeptide. Although it causes synthesis and secretion of both follicle-stimulating hormone (FSH) and luteinizing hormone (LH), it is a much more potent stimulator of the latter.

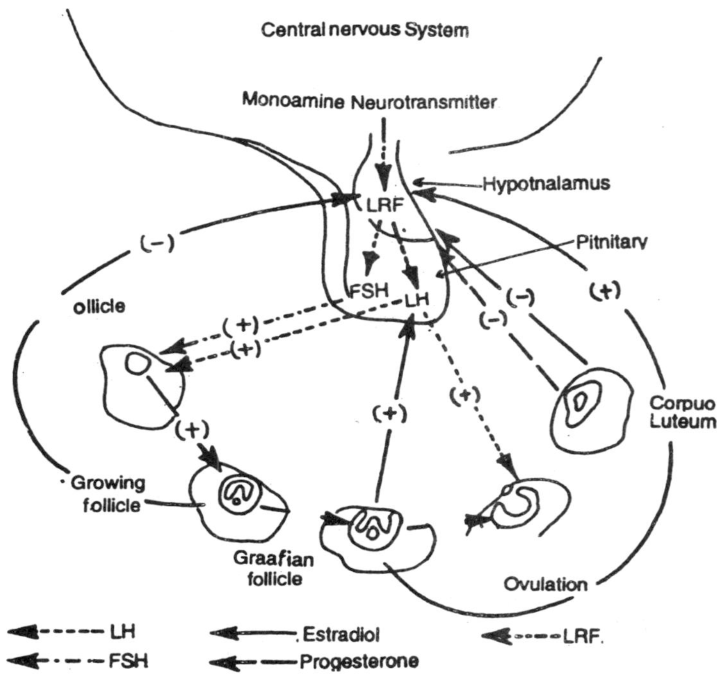

Figure 11.1 : Positive (+) and negative (–) feedback (long-loop) pathways in the central nervous system and the ovary. Arrows point in direction of feedback. See text for explanation of the pathways.

In addition to its inhibitory action on release of gonadotropin releasing factor, dopamine has an inhibitory effect on the synthesis and release of prolactin, probably through synthesis of an as yet uncharacterized prolactin inhibiting factor (PIF).

The LRF neuronal system has been described in the monkey and in the human. The cell bodies of LRF-containing neurons are found in the anterior hypothalamus and in the tuberal hypothalamus. LRF is synthesized in these neurosecretory cells of the hypothalamus, and then released into the hypothalamic-hypohyseal portal system, in which they are transported to the anterior pituitary. At this site, they stimulate synthesis and release of LH and FSH.

The Pituitary

Three gonadotropic hormones of the anterior pituitary have been identified: follicle-stimulating hormone (FSH), luteinizing hormone (LH), and prolactin (PRL).

FSH and LH are glycoprotein hormones containing alpha (a) and beta (a) subunits. They are closely related, structurally and chemically. The a subunits of these hormones are essentially identical. The immunologic specificity of FSH and LH is invested in the B subunit. FSH and LH are synthesized in basophilic cells, called *gonadotropes,* which are scattered throughout the anterior pituitary.

There has been debate about the source of FSH and LH. The critical question was whether both hormones are made in the same gonadotropes, or whether there are two separate populations of gonadotropes, one for LH synthesis and one for FSH synthesis. This question remains incompletely settled, but recent evidence strongly supports the contention that FSH and LH are located in the same gonadotropes in the human pituitary. LRF seems to control both FSH and LH synthesis and release.

The third hormone produced by the pituitary, prolactin, is a single-chain polypeptide composed of 198 amino acid residues. It is synthesized and released by acidophilic cells of the anterior pituitary. The role of PRL in the normal human menstrual cycle has been less precisely defined than the roles of FSH and LH. Recently Vekemans and his coworkers reported a midcycle PRL peak, with continued elevated levels during the luteal phase. However, other investigators have found no cyclic variation in PRL. Current opinion holds that prolactin is necessary for ovarian and testicular steroidogenesis.

FSH and LH exert two types of influence on the reproductive cycle. Their primary effect is stimulation of target cells in the gonad. Although there may be some overlap of activity, because of the structural similarities of the two hormones, each gonadotropin has a specific role in the regulation of gonadal function. FSH stimulates follicular growth and maturation in the female, and initiates spermatogenesis in the male. In addition, FSH catalyses the aromatization of testosterone to estradiol in both sexes. LH stimulates the conversion of cholesterol to pregnenolone in the gonads of both sexes. LH stimulation is indispensable to the complete maturation of germ cells in male and female gonads. However, steroidogenesis proceeds to functional levels under the stimulation of relatively low levels of LH, FSH alone produces neither complete germ cell maturation nor complete steroidogenesis. Thus, it appears that LH, even if FSH is present only at very low levels, can induce steroidogenesis, but that both trophic hormones must be present in adequate amounts, in order for reproduction to be normal. Thus, the gonadotropins are ultimately responsible for the production of the ovarian hormones, which, in turn, control LRF release through the "long-loop" positive or negative feedback system.

In addition to their primary effects on the gonads, the pituitary gonadotropins exert a negative feedback effect on the hypothalamus, through a retrograde "short-loop" vascular system connecting the pituitary and the hypothalamus. Recent animal experiments have demonstrated a neurohypophyseal capillary network common to the median eminence. and the infundibular stalk and neural lobe of the pituitary. Three vascular routes have been demonstrated: fenestrated portal vessels to the anterior lobe of the pituitary; capillary connections to the medial basal hypothalamus, with orientation of the capillary loop toward the arcuate nucleus; and an internal plexus to the ependyma of the median eminence (3,4,17). Additional data have been reported demonstrating retrograde transport to the hypothalamus of LH and prolactin, in the rat. Thus, there is strong evidence for short-loop feedback of the *gonadotropins* to the hypothalamus.

The Ovary

The human menstrual cycle is under ovarian control, in the sense that the ovarian steroids profoundly influence the entire complex. For clarity of exposition, it is convenient to begin with a description of the hormonal events of the menstrual cycle.

The average length of the menstrual cycle is 28 days. By convention, the day of initiation of bleeding is designated day 1. The new follicle for a given cycle starts to develop during the luteal phase of the preceding cycle. Just before and during menses, the cells of the developing follicles proliferate (the follicular phase), and the follicles increase in size, soon to exceed 1 mm in diameter. Many do not attain this size, but become atretic. The follicle that is most sensitive to FSH stimulation gets a lead that it never relinquishes. The rising FSH stimulates the follicle to grow, by inducing mitotic proliferation of the granulosa cells and formation of the theca. The early follicular phase ends when a definite rise in plasma estrogen is noted.

The locally produced estrogen also stimulates the follicle to grow, so that the mature follicle, i.e., the Graafian follicle, reaches a diameter of 6 mm. The next largest follicles become atretic at a size of 1 to 3 mm. The concentrations of estradiol (the major biologically active estrogen) in the general circulation rise from 50 pg./ml. on day 1 to about 75 pg./ml. on day 6 of the cycle. Estradiol concentrations increase more sharply to reach levels of about 150 pg./ml. on day 9. A sharp rise, usually called a "surge," then occurs, to reach a peak of about 3.50 pg./ml. on day 11. Estradiol levels decline rapidly, to values of about 250 pg./ml. on day *14*. The estradiol concentration then gradually rises again.

The central nervous system-hypothalamic-pituitary axis becomes increasingly sensitive to the positive feedback action of estradiol as the concentration of estradiol in the bloodstream increases during follicular development. An acute surge of LH and FSH at mid-cycle is induced by estradiol. This positive feedback requires a circulating level of estradiol of 100 to 200 pg./mL, which is sustained for 36 to 42 hours (8, 9, 23, 26, 29). Current evidence suggests that estradiol is required for the initiation, but not for the maintainence, of the LH surge. In primates, including man, the interval between the estrogen peak and the LH peak is 14 to 27 hours. Ovulation follows the LH peak within 11 to 24 hours.

Shortly before ovulation, plasma levels of 17 hydroxyprogesterone, produced by the ovary, increase. Just after the onset of the LH surge, levels of progesterone, coming from the ovary, also begin to rise.

Around the time of ovulation, the granulosa cells luteinize.

Luteinization is initiated by LH. For the first three postovulatory days (cycle days 15 to 17) the granulosa cells proliferate, and the corpus luteum reaches a size of 1 cm. During postovulatory days 4 to 9 (cycle day 18 to 23) capillaries, previously found only in the thecal cells, grow to reach the granulosa, and peak vascularization begins. After day 23, regression begins. The highest levels of plasma progesterone are reached on cycle days 18 to 23.

To summarize, the following endocrinologic events are hallmarks of the normal menstrual cycle. The estrogen peak precedes the LH peak- 2) estrogen secretion attains appropriate levels; 3) the LH peak occurs at least 13 days prior to the onset of menses; 4) initial rises in plasma 17-hydroxy-progesterone and progesterone levels occur, coincident with the LH rise;-5) a second rise in plasma 17-hydroxyprogesterone occurs later, coincident with the marked rise in plasma progesterone; 6) plasma progesterone begins to rise with the LH surge, and reaches a maximum 6 to 8 days after the LH peak.

The Endometrium

The cyclic changes in the ovary elaborating first estro... and then estrogen and progesterone, induce, in the endometrium, first, proliferation, and then differentiation. These changes are reflected in the endometrial morphology, and the functional state of the ovaries can be inferred with reasonable accuracy from the histologic appearance of the endometrium.

During the first half of the cycle, endometrial glandular and stromal cells undergo replication. The endometrium thickens, and the glands begin to elongate. The epithelial cells become taller, and mitotic activity is seen in glands and stroma. The firstmorphologic change after ovulation is the formation of subnuclear vacuoles in the glandular epithelium. This is first seen on the 15th to 16th day of a 28 day cycle. The vacuoles are caused by the accumulation of glycogen, which on the 19th or 20th day is deposited in the glandular lumen. From the 19th to the 23rd day, no mitoses are seen in either the glands or the stroma, suggesting that progesterone causes cessation of hyperplastic growth. The earliest stromal change seen under progesterone influence is progressive loosening of the stroma, with accumulation of fluid between the cells. This is due to a change in the ground substance, presumably related to preparation for implantation. At about the 23rd day, evidence that progesterone causes hypertrophic growth and differentiation can be inferred from the appearance of changes in the stromal cells surrounding the blood vessels which will allow for implantation. These decidual changes progress over the next few days, until the stroma has undergone generalized decidual transformation. If pregnancy does not occur, the stroma becomes infiltrated with inflammatory cells.

The premenstrual endometrium consists of thousands of tiny units composed of large numbers of stromal cells and exhausted glands surrounding markedly coiled arterioles. The first event in menstruation is vasoconstriction of these arterioles. The resulting ischemia results in dissolution of the architecture of the endometrium. Foci of swelling and distortion appear, the stromal cells begin to clump, hemorrhage occurs within the stroma, and the endometrium is shed as the menstrual discharge or flow. The raw, denuded surface of the endometrium is rapidly covered by epithelium, so that within three days healing is complete.

Thus, the changes caused by progesterone in a non-fertile cycle are peak secretion, followed by peak edema in the endometrium, followed by decidual formation: If conception occurs, instead of each of these endometrial changes peaking and waning in sequence, all continue and peak together. This lends a characteristic appearance to the endometrium, which has been described as "gestational hyperplasia."

INTEGRATION

Integration of the components of the endocrine system is controlled by the hormones described in the first section of this chapter. The same hormones are synthesized by both men and women, but the patterns of secretion are different in the two sexes.

The gonadotropins are secreted in pulses, at intervals of about 90 minutes. Pulse amplitude is greater for LH than for FSH. In males, the pulses are maintained in a tonic pattern, whereas in females the cyclic gonadotropic peak that occurs before ovulation is superimposed upon the tonic pulsatile pattern.

Tonic secretion of LH and FSH is regulated by a feedback loop involving central nervous system components (i.e., dopamine, norepinephrine, and LRF), ovarian steroids, and the gonadotropins. Estradiol is the most potent gonadotropic inhibitor. A quantitative relationship between the negative feedback action of ovarian steroids and gonadotropin release can be demonstrated by interrupting the negative feedback loop by withdrawing estradiol through ovariectomy; this results in significant increases in circulating LH and FSH concentration. The rise in gonadotropins continues until a plateau is attained at approximately ten times preoperative levels about 3 weeks after the operation. The increase is greater for FSH than for LH. This differential effect may reflect preferential inhibition of FSH by estradiol (25, 27, 28) and possibly, by follicular inhibin (6).

There is some evidence that inhibin, a non-steroidal gonadal factor, is involved in the feedback control of gonadotropic secretion. A protein which inhibits FSH secretion has been identified in the testes and in testicular secretions (1,7,11). Because the increase in FSH levels in menopausal women is disproportionate to the increase in LH levels, an inhibin factor has also been postulated to be present in women. Some evidence for the presence of inhibin in ovarian follicles has been obtained in animals (6, 12, 22). Although incomplete, the inhibin story may prove to be an exciting breakthrough, because the presence of a specific inhibitor of FSH secretion and release would explain the differential levels of FSH and LH observed under certain circumstances.

Other observations provide further evidence of the feedback effects of estradiol on gonadotropin output: 1) during the first week after ovariectomy, a greater rise in LH and FSH is seen in women operated upon during the follicular phase of the cycle than in those operated upon during the luteal phase; 2) withdrawal of estradiol results in elevated secretion of hypothalamic LRF and thus of pituitary gonadotropins (5, 10); 3) there is also a rapid decline of circulating gonadotropin levels in response to estradiol administration to ovariectomized and postmenopausal women.

Moderate levels of estradiol, as seen in the early follicular phase of the menstrual cycle, maximally inhibit gonadotropin output. In the normal menstrual cycle, a change in estradiol levels in either direction reduces inhibition and stimulates the hypothalamic-hypophyseal axis. As mentioned above when the hypothalamic-pituitary axis is maximally stimulated through ovariectomy or the menopause, then the action of estradiol will be solely inhibitory. The negative and positive feedback actions are not interrupted phenomena, but rather a continuum. The positive feedback effect of estrogen on gonadotropin release is always preceded by a phase of negative feedback.

To summarize, in response to a nadir in circulating levels of estrogen, gonadotropin releasing hormone is secreted by the hypothalamus. The pituitary, in turn, releases FSH, which stimulates follicular growth. LH, secreted by the pituitary in basal amounts, stimulates synthesis of estradiol by the theca interna of the developing follicles. Estradiol synthesized by the Graafian follicle has a local trophic effect on the follicle and an inhibitory effect on pituitary release of FSH. AsTSH declines, the remaining follicles regress and become atretic. Estradiol synthesis in the ovulatory follicle increases dramatically. Peak levels of estradiol are reached approximately 2 days prior to ovulation. In response to the sustained estradiol peak, a surge of stored LH is released from the anterior pituitary, presumably due to increased secretion of gonadotropin releasing factor by the hypothalamus. A smaller increase in FSH concentration is noted at the same time. Thus estradiol inhibits the hypothalamus in the follicular phase, but stimulates the hypothalamus at mid-cycle.

After ovulation, the granulosa cells become vascularized as the corpus luteum forms. Basal levels of LH stimulate secretion of estradiol and progesterone by the corpus luteum. In combination, these steroids are potent inhibitors of the synthesis of gonadotropin releasing factors. Progesterone may also, inhibit further follicular maturation. In the absence of a luteotrophic factor (hCG), the corpus luteum ceases to function after 12 to 14 days. The cells of the corpus luteum show evidence of degeneration. Although the yellow colour is maintained for months, the structure becomes smaller and is eventually replaced by hyaline tissue. The resultant structure is called a corpus albicans.

two secondary spermatocytes. Each secondary spermatacyte, in turn, undergoes reduction division to produce two spermatids. Development of head, midpiece, and tail then occurs, resulting in mature spermatozoa.

Spermatozoa are released into the lumen of the seminiferous tubuless and transported to the epididymis, where they undergo further maturation and storage. At ejaculation, they are propelled as a bolus through the vas deferens. Seminal plasma secreted by the seminal vesicles and prostate is added to the bolus of sperm as it traverses the prostatic urethra, thus completing the composition of the ejaculate.

As noted above, the testis produces testosterone and other androgens, which are secreted in pulsatile fashion. The testis also produces estrogen, which may be involved in the feedback 'mechanisms at the hypothalamic and pituitary levels.

Thus FSH is important for spermatogenesis and LH for production of testosterone. The regulation of the components of the hypothalamic-pituitary-testis system is similar to that in the female, with testosterone and testicular inhibin playing major roles in the long-loop feedback system. The short-loop system of pituitary tropic hormones (FSH and LH) feeding back on the hypothalamus and its releasing hormone (LRF) is also probably operative. Since the production of LH in the male is only tonic, and not both tonic and cyclic as in the female, the regulatory system is simpler and less susceptible to disruption.

Testosterone has two major roles in the male. It stimulates development of male secondary sex characteristics, such as enlargement of the genitalia, increase in laryngeal size with concomitant deepening of the voice, increase in muscle mass, and growth of body hair. It also is indispensable to normal spermatogenesis. The effects of androgens on other tissues and organs are discussed later in this chapter.

The Ovary

The fetal ovary does not demonstrate the hormonal activity seen in the fetal testis. Ovarian steroid synthesis is not necessary for differentiation of female external or internal genitalia. However, the fetal ovary actively produces germ cells. The full complement of ova is achieved by 20 weeks of intrauterine life, and the process of follicular maturation and regression is evident by the third trimester of pregnancy.

The ovary is composed of numerous follicles, in varying stages of maturation, surrounded by stromal cells. Follicular units consist of an ovum surrounded by a layer of granulosa cells, and internal and external layers of thecal cells. The maximum number of follicles (millions) is reached at approximately 20 weeks of intrauterine life. From this time, follicles become involved in an inexorable cycle of partial maturation followed by atresia. Most follicles become atretic; only about *500* develop to the stage of ovulation in a lifetime.

Gonadotropin levels in females are low in infancy and rise gradually as puberty approaches. Levels of FSH are slightly higher than those of LH, but the ratio is nearly one to one. In contrast,, during reproductive life LH is always higher than FSH, except early in the follicular phase of the cycle, when growth of a new crop of follicles is stimulated.

The rising gonadotropin levels of late childhood stimulate estrogen synthesis in the ovarian stroma. The secreted estrogen causes maturation of the external genitalia and stimulates growth of the breasts. Eventually, when there has been sufficient estrogen stimulation of the endometrium, uterine bleeding occurs. The first episode of bleeding is called the menarche. The initial endometrial shedding probably occurs in response to a drop in circulating levels of estrogen. The first several "menstrual" episodes may occur at irregular intervals because ovulation has not yet been initiated. Once an ovulatory pattern has been established, it recurs with consistent periodicity in an individual woman. Toward the end of the reproductive age, ovulation and menstrual cycles again become irregular. At menopause, estradiol synthesis decreases drastically, and gonadotropin levels become markedly elevated.

The pattern of ovarian steroid secretion fluctuates throughout the menstrual cycle. The ovulatory follicle and corpus luteum are the major sources of estradiol synthesis and secretion. Average production rates range from 0.07 mg./day in the early follicular phase to 0.8 mg. /day at the time of the preovulatory peak. Estrone is also secreted by the ovary, but most of the circulating estrone is derived from peripheral conversion ofandrostenedione. About half of circulating androstenedionee of ovarian origin and half is of adrenal origin; it is secreted by the ovary in a cyclic pattern which, parallels that of estradiol and estrone. Mean testosterone levels in the normal woman are about 40 ng./ml. of serum. Approximately half is secreted by the ovary and the remainder by peripheral conversion of androstenedione and of dehydroepiandrosterone, an androgen mostly produced (in normal circumstances) in the adrenal gland.

About 50 per cent of the estrone and estradiol secreted daily undergoes 16-hydroxylation to form estriol. Estriol is then conjugated with sulfate or glucuronide to facilitate excretion.

Progesterone and 17-hydroxyprogesterone are predominantly secreted by the corpus luteum during the luteal phase of the cycle. Low levels of progesterone found in the follicular phase of the cycle are presumably of adrenal origin. As noted earlier, ovarian progesterone and 17-hydroxyprogesterone secretion become detectable in the periovulatory period concomitant with the LH surge. Peak secretion of progesterone and 17-hydroxyprogesterone in the mid-luteal phase is about 24 mg./24 hours and 4 mg./24 hours, respectively. Progesterone gradually rises, beginning in the periovulatory period, whereas 17-hydroxyprogesterone rises sharply before ovulation, and again during the midluteal phase.

Mechanisms of Hormone Action

Gonadotropins and steroids select their target cells because of the presence of specific receptors on, or in, these cells. The mechanism of hormone receptor interaction differs for each class of hormones.

The protein hormones, FSH and LH, attach to receptors located on the plasma membrane. The resulting membrane receptor-hormone complex, in turn, activates a membranebound enzyme, adenylcyclase. In the presence of intracellular magnesium, adenylcyclase acts on ATP to produce cyclic AMP. Cyclic AMP then activates protein kinases, which mediate the biologic responses of the cell. In the case of LH, steroidogenesis is initiated, and in the case of FSH, specific proteins are elaborated which affect germ cell growth and development.

In contrast, steroid hormones combine with specific receptors in the cytoplasm of target cells and are carried by these receptors to the nucleus of the cells, where they interact with specific sequences of DNA to effect messenger RNA synthesis and protein synthesis.

Steroids secreted by gonads are transported to target tissues in the bloodstream in free and bound forms. Testosterone and estradiol bind to a specific binding globulin known as TEBG (testosterone-estradiol binding globulin) or sex steroid binding globulin. Progesterone has no specific binding globulin of its own but appears to bind preferentially to corticosteroid binding globulin (CBG). Only free steroids are biologically active in target organs and in the feedback mechanism.

In addition to feedback relationships with the hypothalamus and pituitary, gonadal steroids affect a variety of target organs including the skin and its appendages, subcutaneous fat, muscle, breast, larynx; uterus, cervix, vaginal mucosa, external genitalia, and the gonad itself.

Androgens exert their primary tropic effect on skin, muscle, laryngeal tissue, and male external genitalia. Androgens are responsible for the laryngeal enlargement which produces the deep male voice and for the increased muscle mass characteristic of the male. Androgens stimulate penile and testicular growth and are responsible for rogations of the scrotum. They stimulate growth of body hair and secretory activity of sebaceous glands and, paradoxically, are also responsible for the temporal recession of the hair line seen in males. At the peripheral level, the aetive androgen is 5a-dihydrotestosterone, which is usually derived by peripheral conversion of testosterone through an irreversible reaction mediated by 5a-reductase.

Estrogens are responsible for growth of the breast and development of the ductile system, and for the deposition of fat *in* the breasts, abdomen, hips, and things which produces the characteristic female body habitus. Estrogens also induce hyperplastic growth of the endometrium, increased production of cervical mucus, and thickening and rugation of the vaginal mucosa. Estrogen induces the growth of estrogen and progestrone receptors in target tissues, whereas progesterone eliminates these receptors.

Progesterone is a complement, as well as an antagonist, to estrogen. In the breast progesterone induces growth of the lobule-alveolar complexes after estrogen-induced growth of the ductile system has taken place. Estrogen causes increased water content of cervical mucus, resulting in a copious, clear mucoid secretion, and progesterone induces increased viscosity and production of an opaque, gummy mucoid material.

The endometrium is probably the single most important tissue that the responsive to the effects of both estrogen and progesterone. Estrogen induces mitotic growth of the endometrial glands and stroma, and increased endometrial blood flow. Progesterone halts proliferation of endometrial tissue and induces hypertrophic growth and differentiation of the existing tissues. The interaction of these two steroids is finally responsible for the cyclic bleeding referred to as "menstrual periods."

Gonadal steroids also affect target tissues that. are not related to sexual or reproductive function. The anabolic action of testosterone results in positive nitrogen balance and retention of potassium, phosphorus and calcium. Testosterone also stimulates hemoglobin synthesis. Both testosterone and estradiol are vasoactive and abrupt depletion, as experienced in loss of gonadal function, produces vasomotor instability, with accompanying "hot flashes."

Estradiol increases serum levels of trigly-cerides and hormone binding globulins, including those for sex steroids, cortisol, progesterone and thyroxine. Estrogen increases insulin response to carbohydrate and causes sodium retention, through interaction with the renin-angiotensin system; it also affects the ability of the liver to secrete substances such as bromosulfophthalein. Progesterone induces natriurisis, probably by competing with estradiol.

All three major sex steroids may cause changes in mood. Estradiol and testosterone are alleged to enhance libido and contribute to a feeling of optimism and well-being. Progesterone is said to contribute to depression. Evidence for these effects is largely inferential, and requires further documentation.

Index

D

E

F

G

H

I

J

K

L

M

R

S

X

Y

Z